COLLECTIBLE PLASTIC

Kitchenware and Dinnerware

1935–1965

PLASTICS BOOM

MICHAEL J. GOLDBERG

Schiffer Publishing Ltd

77 Lower Valley Road, Atglen, PA 19310

FOR FURTHER INFORMATION

If you like plastic kitchenware or dinnerware from this period and need further information on identifying pieces or pricing, please contact the author, Michael J. Goldberg through the publisher.

Printed in China
ISBN: 0-88740-843-5

We are interested in hearing from authors with book ideas on related topics.

Library of Congress Cataloging-in Publication-Data

Goldberg, Michael J. (Michael Jay)
    Collectible plastic kitchenware and dinnerware, 1935-1965 / Michael J. Goldberg.
    P.   cm.
    Includes bibliographical references and index.
    ISBN: 0-88740-543-5 (pbk.)
    1. Plastic tableware--Collectors and collecting--United States--Catalogs. 2. Plastic kitchen utensils--Collectors and collecting--United States--Catalogs. I. Title.
    NK8595.G66   1995
    668.4'9--dc20   95-15968
                                                     CIP

Published by Schiffer Publishing Ltd.
77 Lower Valley Road
Atglen, PA 19310
Please write for a free catalog.
This book may be purchased from the publisher.
Please include $2.95 postage.
Try your bookstore first.

Contents

"Promoted to the head of its class! — dinnerware molded of MELMAC Plastic." From *Modern Plastics Magazine*, 1946. *Photography by Dixon.*

Acknowledgments

For Standard Plastics Company by PH. From *Modern Plastics Magazine*, 1957.

Many, many thanks to the following:

Jim Sutherland for his incredible patience and support.

Cat Yronwode without whose encouragement this book may not have come into being.

Chuck Meyer of the Badger Club for his photographic assistance.

Florence T. Crowell of the Watertown Historical Society, Watertown, CT.

Elise Feely of the Forbes Library, Northampton, MA.

Al Lavoie of English and English Inc., Boontonware Dinnerware. Bloomingdale, NJ.

Charlie Nelson of Palookaville, Portland, OR.

Mike O'Hara of Modern Times, Portland, OR.

David and Claire of the Old Town Antique Market, Portland, OR.

Periodicals Paradise of Portland, OR.

Steve, Don and Loretta of the St. Vincent DePaul, Portland, OR.

Janis at the San Francisco Public Library, San Francisco, CA, and David King, Darryl Ballini, Rose Roopenian, Eliza Kruck, and Dennis Wanken.

My editor Leslie Bockol and publisher Peter Schiffer for helping to make this book a reality.

Advertisement for Emeloid kitchen wares. From *Modern Plastics Magazine*, 1946.

Chapter One
An Introduction To Plastic Kitchenware and Dinnerware

In 1883, Laura C. Holloway, a renowned author of her times, stated in her book *Life at Home, A Household Manual*: ". . . as every human life depends upon food for its support, so every human home must have a kitchen in which that food is prepared. The person who neglects food soon gets sick, and the home in which the kitchen is neglected is not a healthy or a happy home."

Though health and happiness are not as easy to secure these days, the incredible supply of housewares made for the kitchen indicates this room has been anything but neglected. The kitchen has undergone more technological, utilitarian and decorative changes in the twentieth century than any other room in the home. In the wake of all these changes, the kitchen has left us a multitude of items — items particular to the times in which they were created. It is no small wonder that these 'kitchen collectibles' have become an area of popular interest in the twentieth century. Wooden butter churns, enameled cookware, Hoosier cabinets, apple corers, cast iron skillets — all these and much more have been cherished and sought after by kitchenware collectors for years. Even forty years ago, when 'antique' meant Chippendale dressers, kitchen items like wooden coffee grinders and tin molds were considered "charmingly old fashioned" and were in demand. Many of these desirable items date from the 1900s all the way back to colonial times. Interest in kitchen collectibles from this century has taken a long while in coming into its own. People are now collecting items not originally considered 'antique' — from Depression era glassware to wartime electrical appliances and post-war chrome dinette sets — with a passion and a new appreciation.

One of the things that separates kitchenware and dinnerware of this century from prior ones is the introduction of plastics into the kitchen. The 1930s and 1940s saw the earliest plastic kitchenware items, and that field is already popular with collectors. The domestic use of plastic had some of its greatest successes in the kitchen during the post-war years, and these items are suddenly generating a new interest as well. By the late 1960s almost everything in the up-to-date kitchen could be made (or partially made) of plastic — not just dishes but countertops, floors, wall coverings, storage bins, furniture, utensils and gadgets. Combine all of these eras of plastic kitchenware and dinnerware, and you have a vast field to collect from.

Now is a great time to start a collection. With many people not aware of plastic's collectibility and the knee-jerk human response to regard plastic as cheap and dismissable, the field is wide open. 'Antique' kitchenware from past centuries, as well as the highly popular ceramic dinnerwares of the 1930s through the 1950s are disappearing rapidly — not only into collections but into museums as well. Kitchen items from the early part of this century are still around, but their supplies are slowly dwindling, too. As we approach the end of this century we are far enough away from the post-war years to appreciate the products of that time as collectible, if not antique. It is hard for many people to consider plastic as old, but remember, plastics have been around for almost a hundred years. The 1930s, 1940s and 1950s were the early years of plastics in regard to kitchen items. Some plastics used then are now obsolete in their original form. Most are still in use (even Bakelite!) but have been technologically updated according to modern discoveries. The colors and styles have also changed tremendously. With plastics such a large part of our environment today, these early plastic kitchen items seem — dare I use the word? – *quaint*. Someday, these too, may be in museums.

This book is a result of months and months of researching and years of collecting. In a somewhat Christopher Columbus-like fashion, I have charted the seas of 'collectible' plastics with very little info. The waters of doubt and disbelief have been choppy ("You collect WHAT?"; "Old Plastic? — give me a break!"). I have kept steady on my course and this book is a result of that long journey. All the pieces shown in this book are from my own collection unless otherwise noted.

Finally, I hope you enjoy reading and using this book and find it a helpful source with which to explore this new field of plastic collectibles. Please read the section on identifying the plastics shown in this book. *All plastic is not the same.* Learning the differences will help in your collecting, so shake off your prejudices about plastic and examine the possibilities. You'll be surprised.

TYPES OF COLLECTIBLE PLASTICS

Most of the 'modern' plastics in this book were developed in the 1920s and 1930s, put to war production in the 1940s and came to full fruition in the post-war years. From the early 1930s until the end of the war, plastics were beginning to be introduced into the kitchen. The plastics available at this time, though tough and revolutionary, were not entirely suited for the rigors of cooking and cleaning. Two plastics, *Bakelite* and *Catalin*, were used for handles, knobs, and utensils. *Cellulose Acetate* and *Urea*, two other early plastics, found use in bowls and dishes (though not very successfully) as well as appliance housings and hundreds of kitchen items. However, three plastics were being developed that would revolutionize the kitchen. After the war, *Polystyrene* and *Polyethylene* were discharged into civilian life only to be signed up for permanent kitchen duty. These two tough plastics took on every possible kitchen item — and won! *Melamine*, the third plastic, went on to become the star Melmac, of plastic dinnerware fame. (See dinnerware section.)

Advertisement for a bread box and matching canisters made by the Plas-Tex Corporation. The Polyethylene canisters are described as "squeeze bottle" plastic. From *Good Housekeeping Magazine*, 1954.

Advertisement for Lustrex Plastic. Lustrex is a trade name for Polystyrene produced by Monsanto Chemical Company. From *Good Housekeeping Magazine*, 1954.

Below is a listing of the 'modern' plastics you will encounter in this book. You might recognize some of them already. If you don't — that's O.K. Many plastics are hard to identify by look and feel, especially the older ones. It's important that you learn to identify these plastics.

Bakelite — This is the first entirely man-made plastic. Developed in 1907, Bakelite is the brand name for what is really any molded phenolic resin. It's main 'plus' is that it is highly heat resistant. On the down side, Bakelite's base resin was dark amber — it could not be made in bright or pastel colors or white. Black, dark browns and burgundy are the most recognizable colors of Bakelite items. You will find Bakelite mostly as fixtures on appliances and other pieces — knobs, handles, radios and telephone bodies are instantly recognizable Bakelite parts. Also to its disadvantage, Bakelite yellows with age and becomes dull and brittle. You will find some smaller Bakelite kitchen items but not many.

Catalin — Catalin is, chemically speaking, Bakelite's first cousin. Catalin is a *cast* phenolic resin (Bakelite is a *molded* phenolic resin) with a clear resin base. This made it superior to Bakelite, since it could be produced in a wide variety of colors and with marbelized, translucent, and two-toned effects. The great majority of pre-war and wartime kitchen items pictured in this book are made of Catalin, particularly the handles on flatware and utensils. Sadly, Catalin became increasingly expensive to make and could not compete with the post-war plastics that were less expensive to manufacture. Catalin's beautiful colors tend to fade rapidly if exposed to light.

Acrylic — Developed around 1927, acrylic has been used mostly for its transparent qualities. Its use as a substitute for window glass has made acrylic instantly recognizable to most people. It is also odorless, lightweight but tough, and can be frosted, etched or painted. Older acrylic becomes brittle, yellows rapidly with age, and scratches easily. These three unfortunate traits, however, make older acrylic easy to identify. Serving pieces, tumblers, pitchers, utensils, appliances and vases are but a few items you will find in acrylic. Do not confuse acrylic with clear Polystyrene (see Polystyrene).

Cellulose Acetate — Also developed in the late 1920s as a sturdier and better quality Celluloid (Cellulose *Nitrate*), this lesser-known plastic was actually the first to make inroads into the plastic housewares arena. Tumblers, trays, soap dishes, shakers, and utensil handles were a few of the many items made of Cellulose Acetate. Despite its many uses, this popular plastic was not heatproof and tended to warp easily (though it was not flammable like Celluloid). It also became brittle. The fall of Cellulose Acetate paralleled the rise of Polystyrene. Some of the earliest kitchen items are made of Cellulose Acetate. They are neither easily found nor plentiful. Cellulose Acetate is sometimes hard to identify. It looks and feels like a cross between Celluloid and Polystyrene.

Urea-Formaldehyde — Developed by the British in 1924, Urea-Formaldehyde was a transparent resin which could be made in colors as well as white. In the early 1930s, the American Cyanamid Company bought the rights to produce Urea and greatly improved it. The British Urea dinner pieces had been marketed under the names of Beetleware and Bandalasta. They were sold in bright colors or with a marbleized effect. The American-made Urea of the 1930s and 1940s was also sold under the name Beetleware and was used on novelties, picnic items, lunch sets and children's items. The juvenile pieces are the best-remembered. Little Orphan Annie mugs and countless other tumblers, cereal bowls, and feeding sets were offered. The pieces are thick and hard. Not much Urea has survived, as pieces made from this plastic swelled and cracked when placed in water. A chemically improved version of Urea is still produced today, and is used in such things as electrical and mechanical housings.

FEAR OF MELAMINE

In 1946, The Boston Police Department received a local report that Melamine, if burned, would give off a deadly cyanide gas. Melamine had been used extensively for military purposes during the war, and by 1946 was being used more and more by civilians for the interiors of buildings (laminates, tiling, and housings) as well as for dinnerware. Still, civilians were not as familiar with the advances made in plastics as the Armed Forces were, and this new report caused quite a stir.

With war stories of tear gas, smoke bombs and other chemical warfare terrors still in the public mind, the Boston Police Department's worry turned to widespread panic. Fire departments across the country called for laws to control the use of plastics. The tension spread to the local citizenry. Newspapers reported that housewives thought to have dropped dead from heart trouble may actually have suffered from cyanide poisoning caused by fumes from overheated Melamine dishes.

No, housewives were not running out in droves to their local dumps, arms filled with Melmac; in reality, the initial report had been spotty and inconclusive. Chemist Foster F. Snell went to Boston and cleared up the questions, demonstrating with rats to put an end to the rumors. Yes, cyanide gas could be deadly -- but not from breathing the fumes, except in massive doses. He also explained that any nitrogenous item (including wool, beef, and leather) gave off small amounts of cyanide gas when heated or burned. Mr. Snell noted that the average housewife would be astounded to know that such common substances as newsprint, steak, milk, and clothing produce toxic gas if burned under certain conditions. Mr. Snell concluded on an optimistic note, assuring firemen that new products like plastic were less of a fire hazard than the more familiar products they had superseded.

Melamine — First 'discovered' in 1834 in Germany, this plastic powder was considered an oddity. However, in 1937, scientists at American Cyanamid Corporation rediscovered Melamine in their search for a sturdy plastic for wartime use. Produced for many different purposes during the war (helmet linings, machine casings,) its greatest use was for dinnerware for the armed forces. Its toughness and durability were well suited for such application. To this day, no other plastic has replaced Melamine for dinnerware. Though other companies produced Melamine, American Cyanamid called their product 'Melmac', now the widely-known generic name used to describe any piece of plastic dinnerware made of Melamine. Melamine is easy to recognize. It is a tough, glossy, hard plastic that comes in wide range of colors as well as mottled, speckled and even clear.

TUPPERWARE FOREVER

No book on the history of plastic kitchenware would be complete without a mention of The Tupper Corporation. In 1938, Earl S. Tupper started his business. His first product was produced at the end of the war — a Polyethylene tumbler in frosted white. A year later, the line was expanded to include color. Soon bowls appeared with the now-famous Tupperware Seal, tight-fitting lids made possible by the use of Polyethylene. This was a real 'first' and quickly captured the kitchenwares market. Originally called Poly-T, this line of Polyethylene products was now called the Tupper Millionaire Line. At this time the first home party plan was tried and became so successful that by 1951 Tupperware Home Parties Incorporated was formed and retail store sales were discontinued.

The number of different Tupperware items produced is mind-boggling. But is it collectible? A tremendous amount of Tupperware exists (and it is still being produced), and the colors and styling have changed little over the last 25 years. Because of this, antique pieces may not be worth any more than new versions or used thrift-shop pieces. Still, I feel that right now some of the older pieces are worth collecting. You will need to refer to backstamps. Any Tupperware piece marked "Poly-T," "The Tupper Corp.," "Tupper!" or "Tupper Millionaire Line" are from the original sets. Surprisingly, very few of these pieces have surfaced.

Antique Tupperware? You never can tell!

Polyethylene — Known by many as the "Tupperware Plastic". This "new and revolutionary" plastic (or so the ads called it in 1942) didn't really enter the kitchen in full force until the mid-1950s. It truly was revolutionary — flexible bottles, hinges, ice cube trays and flexible everything else changed the face of kitchenware items forever. By the 1960s, Polyethylene had eclipsed Polystyrene as *the* kitchenware plastic. Its waxy texture and flexibility make Polyethylene easily recognizable. It is extremely tough and resistant to breakage. Polyethylene originally came in a translucent, frosted shade, but eventually could be made in any color. Since most kitchenware items are still being made of Polyethylene, it is easy to confuse the old and the new. Look for backstamps, dated colors and styling. Newer Polyethylene is tougher than the older pieces as technology has greatly improved this all-important plastic. Many companies made Polyethylene items, so not all Polyethylene is Tupperware. Many people were initially repulsed by Polyethylene's early, waxy texture, but attracted by its practicality. Do not confuse Polyethylene with Vinyl — Vinyl is more rubber-like.

Polystyrene — On the eve of World War II, Polystyrene was born as an answer to the United States' loss of its natural rubber supply. As other natural materials went into rationing, Polystyrene proved to be a bigger boon. After the war, Polystyrene kitchen items were released, but not until quality improved in 1950 did they really start selling in earnest. By the 1950s, Polystyrene dominated the kitchenware field. For the first time, kitchenware began to sell in significant volume. You can easily identify this plastic — it is usually lightweight, hard, and has a tinny, almost metallic feel when tapped. It comes in bright, cheerful colors or can be made transparent. Starting in 1951, it was also produced in metallic colors. It is highly chemical- and acid-resistant, waterproof and (for all you manufacturers out there) cheap to produce. On the down side, it warps easily from heat and cracks easily because of its brittle nature. Many kitchen items you see displayed in this book will be made from Polystyrene. Columbus Plastics Co. and their Lustro-Ware line are a good example of the wide range of products made in Polystyrene. As mentioned, cracked pieces are common.

CARE & CLEANING OF PLASTIC KITCHENWARE

A few simple rules will help keep your plastic kitchenware in good shape. In all truthfulness, it is hard to find items without some sort of wear — think of how much use kitchen items get. This fact will also hold true if you plan to use your kitchen pieces daily. Expect some wear and tear to develop.

Bakelite, Catalin, Cellulose Acetate and Urea should *never* be left in water for any lengthy period of time to soak away dirt. These older plastics will absorb water and warp and crack. Clean only in warm water, never hot or boiling. Clean quickly and wipe dry. Bakelite and particularly the beautifully colored Catalin can be washed in a mild ammonia and soapy water solution. To keep Catalin pieces in good shape a product called Simichrome is available. This Brasso-like paste is imported from England and can be purchased primarily at antique stores. Use as you would use Brasso to keep your Catalin pieces shiny and attractive.

Polystyrene is tough but brittle. Be careful in handling pieces made of this plastic as it cracks easily. Always clean with a non-abrasive pad or cleanser as Polystyrene scratches easily, too.

Polyethylene, on the other hand, is tougher. It can take heavier handling and cleaning. It rarely breaks, but pieces such as hinged lids can rip at weak points. Though considered unscratchable, it is wise not to clean Polyethylene with an abrasive pad or cleanser. Though Polyethylene is sometimes advertised as boilable, I wouldn't suggest it on older pieces.

Acrylic is relatively tough. Remember, it scratches easily, so no abrasive pads or cleansers.

In general, a regular washing in warm water with dish soap will clean most pieces of plastic kitchenware. Caked-on dirt and grime are frequently found on old kitchenware items. This can be removed with an all purpose cleaner (a 409 type) and some light but persistent elbow grease. An old toothbrush will help get in odd places. Always rinse your pieces off immediately after cleaning.

Those thrift store sticky-back labels can be removed with a rubber cement thinner (also sold under the brand name Bestine). Soak the label and peel off with your thumb or fingernail. Rub off any remaining glue with a rag or tissue soaked in the thinner. Rinse off immediately. Pieces that crack or break can sometimes be glued back together. I use Crazy Glue. Read the labels as some glues are not meant for plastics and will melt them.

Brighten Her Day with
UNITANE® OR-342
RUTILE TITANIUM DIOXIDE

Advertisement for American Cyanamid coloring agents (in this case white) for plastic products. Maybe she's mixing some up now! From *Modern Plastics Magazine*, 1958.

Plaskon advertisement showing the kitchen of tomorrow, full of the plastic products promised for the post-war era. From *Modern Plastics Magazine*, 1945.

Chapter Two
Kitchenware's Shaky Start and Glorious Finish

Products of PLASKON Plastics and Resins in this kitchen:
Nylon—baby bottles, cabinet knobs, food bag
Urea—electric blender base, radio housing, electrical outlet and switch plate, can opener
Melamine—dinnerware, washing machine agitator, utensil handles
Polyester Resins—translucent panels, chair

PLASKON

Plaskon advertisement 10 years later, showing plastics used in a contemporary kitchen. From *Modern Plastics Magazine*, 1956.

The entrance of plastics into the kitchen (and into the home in general) during the mid-1930s was not an easy one. The first new material in centuries, plastic was given the status of a 'miracle'. The raw material manufacturers had little idea what their products could and could not do. The molders were just as inexperienced. The machinery to create a molded object was not totally sufficient to deal with the new plastics of the time. Everyone in the industry was sort of 'winging it'. To add to the inexperience, many sellers, in their enthusiasm, made incredible claims as to the durability and use of these early plastic products. Though housewives seemed interested in the new, brightly-colored plastic kitchen items, plastics in the kitchen had a shaky future ahead. Inferior items flooded the market — utensils, handles, and cleaning items that easily cracked, swelled or melted. Consumer dissatisfaction rose. At the same time, the country was in the midst of the Depression. Contrary to popular belief, the first kitchen items were not cheap. The initial high cost of producing early plastic kitchenware items made them sometimes more expensive than their competitive markets. In those lean times saving even 50¢ meant a lot for the average family. A good example of this is Rubbermaid's first product, a semi-synthetic rubber dustpan, which cost almost a dollar when introduced in the 1930s. By comparison, a metal dustpan cost 39¢ at the same time. The initial sales re-

sistance because of price was overwhelming. People simply couldn't afford to spend the money. Still, many of the giants of the plastic kitchenware industry (including Columbus Plastics, Beacon, Rubbermaid, and Tupperware) got their start during these tough years though. These companies believed that eventually better quality kitchenwares could be produced in plastic.

It was not until after World War II that the plastics industry really got going, despite another shaky period due to grand promises and overzealous expectations. Like the manufacturers of many items unavailable during the war, the plastics industry made promises about "after the war. . . ." In the case of kitchenware, the Kitchen of Tomorrow (not the future, just tomorrow) was a favorite subject; all kinds of plastic gadgets, appliances, handles, etc. would fill this kitchen. When these items did flood into American homes at the end of the war, problems of quality and consistency arose. Though the earlier advances in plastics had worked technically and aesthetically for war and institutional use, many items were not ready for consumer use. The post-war plastics Polystyrene, Polyethylene, and Melamine were entering the consumer market for the first time. Polystyrene in particular was highly popular among manufacturers because there was a surplus of it around (remember the rubber shortage?), and they quickly put the material to use. But when the first barrage hit the stores, retailers and customers complained of imperfections in molding, streaking, limited color range, brittleness and various other deficiencies.

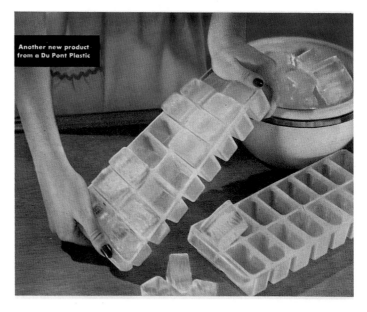

Want to make a hit with the little lady in the apron? Then give her kitchen-ware of plastics . . . like these products molded by us for Devine Foods, Inc., Chicago.

No wonder the lady is all smiles! Those handy mixing bowls can be jostled into crowded refrigerators without breaking or chipping. Their super-smooth surface cleans easily, won't tarnish, and will not impart a foreign taste to foods. The bowls nest handily inside one another . . . with the covers on. And they will even stand oven temperatures required for baking custards.

The two trays are designed especially for cafeteria use. Food is served directly into the compartmented tray to save the bother and expense of dishes. At present the entire production of this plastic kitchen-ware is being taken by the armed forces . . . principally for use in hospitals. After Victory, however, it will find a ready and eager civilian demand.

The commercial success of this plastic kitchenware depends largely on molding to exact standards . . . an assignment which Devine Foods, Inc., entrusted entirely to us. We'll be glad to assist your engineers in developing similarly successful products . . . or will submit quotations based on your present specifications. MOLDED PRODUCTS COMPANY, 4533 W. Harrison St., Chicago (24) Ill.

MOLDED PRODUCTS

Plastics have a date in the post-war kitchen. Devine Foods, Inc. produced an enormous percentage of the early kitchen plastics. Here the war-time housewife is being told "At present the entire production of this plastic kitchenware is being taken by the armed services. After victory, however, it will find a ready and eager civilian demand." The advertisement is for the Molded Products Company of Chicago, IL. From *Modern Plastics Magazine*, 1944.

With the war over, the industry worked hard to bring about technological advances, new plastics, and stricter controls which improved the quality of plastic kitchenwares. By 1947, the industry began to bolster the public's awareness of good and bad quality in plastics through ads and other promotion campaigns. Quality control bureaus were established to set standards within the industry. By the end of the decade, millions (no, make that zillions) of better quality kitchenware items were made, and were immediately bought by an eager public used to wartime scrimping and doing without.

Another factor were the mechanical advances that had been made rapidly since the early days of plastic manufacturing. At the turn of the century, whole new types of machinery had to be built to work with plastic's peculiar physical nature. The early machinery was not well developed, resulting in flaws and problems, inconsistency being a major part. These problems plagued the industry through the post-war years as new plastics were developed — not always concurrent with the machinery and chemistry of the time. It took almost five years, for example, to get the 'bugs' out of some of the post-war plastics back in the late 1940s!

Technologically, plastics were being improved yearly — for example: asbestos was added to Bakelite to improve heat resistance and rubber was added to Polystyrene to counter its brittleness. Color, texture and durability were constantly being updated and improved. By the 1960s, mechanical and technologically advances really took off. The machinery and plants had achieved an incredible state of modern efficiency. No more working out of small brick factories or wooden buildings, as the original companies did. Many plastic companies expanded or updated their facilities in this later post-war period.

In the laboratory, things were brewing. Groups of scientists and researchers now tackled problems formerly left to a small team of chemists (or sometime a sole chemist) back in the early days. These "research teams of scientists and experts," as the copy always called them, hastened development. Plastics had now become such a large part of daily life, the plastics industry did not have to spend continuous energy on convincing the retailers of plastic's legitimacy. Likewise the retailer didn't have to spend money and time selling plastics to a consumer public, now familiar with plastic's qualities. In fact, the word "PLASTIC" in ads, once large and commanding, had slowly gotten smaller and smaller. By the mid-1960s, some companies selling plastic products referred to them as "synthetics" or "chemical made."

Ice pops out at the twist of a wrist. Flexible ice trays, made of Du Pont Polythene, were among the many new products "going plastic" during the post-war period. From *Modern Plastics Magazine*, 1947.

By 1970 almost forty different plastics were in use in housewares! Take a look around your own kitchen and you will see how many items are fully or at least partially made of plastics. Many kitchen items had been made of the same materials for years. At the peak of post-war productions, their conversions to plastic became complete. Take, for example, the spatula. A 1930 spatula was made of metal with a wooden handle. The metal body was screwed or bolted into the handle and the handle itself was painted (green or red being the favorite colors). In 1940 the wooden handle was replaced with one made of plastic (probably Catalin). The metal body and the plastic handle could be fused to insure unbreakability, and the brightly colored handle replaced the painted wooden one — no more chipping paint. By 1970, all-plastic spatulas existed.

The wonders of high-density Polyethylene brought to the kitchen, with new unbreakable plastic housewares made of Marlex. From *Good Housekeeping Magazine*, 1959.

Spatulas through the ages. (Left to right) Circa 1935, 1945, 1955, 1965.

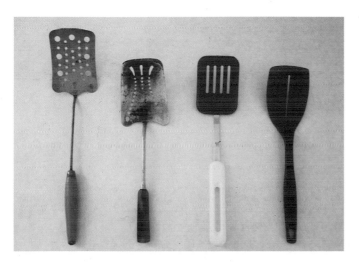

COLLECTING PLASTIC KITCHENWARE

So where does one begin to find this 'old' plastic? Much has survived, but little of it in perfect condition. Still, with the vast quantities produced, finding enough pieces of older plastic kitchenware is no problem. Where to look? First, try your parents' or grandparents' basement, garage, or attic. Check the cupboards, too. Unless a piece was severely damaged, some housewives simply relegated old pieces to back shelves and storage areas. You might even have some pieces in your own home inherited from Mom when you first left the nest. Finding pieces on the homefront is great because, well, they're free!

The best places to buy pieces are thrift stores, garage sales and church bazaars. Right now, many thrift stores have plastic sections where all plastics are thrown together. You will find many interesting pieces here. Garage sales and church bazaars are also great places for good finds because most people are glad to get rid of that 'ugly old plastic stuff' and can't imagine why anyone would want it. (I mean — with all that brand new Tupperware out there!) Don't bother with antique auctions — they're only interested in antiques. A caution about shopping in thrift stores — many newer items are being made of plastic. *Read the backstamps!* Also, pieces of things, particularly parts of childrens toys and modern appliances separated from the original complete item, can sometimes look like a piece of older plastic kitchenware. Be careful, and again, *look for backstamps!*

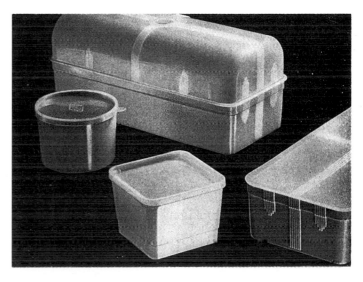

Housewares made of Alathon. This 1957 entry into the polyethylene parade was "attractive due to its high gloss."

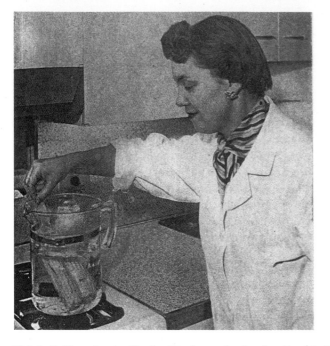

This boilable polyester film bag for frozen foods advertised in *Modern Plastics Magazine*, highlighted the 'heat and eat' trend in 1956. *Photography courtesy of Standard Packaging.*

There is a new breed of 'antique' store appearing in most cities and towns. I call them Collectible Stores. They primarily sell items from the 1930s through the 1970s and some even into the 1980s! Many of the items are from the 1940s, 1950s, and 1960s. Most of the shop owners are baby boom age or younger. These people view the artifacts from these eras with a whole new appreciation. In addition, there are stores called 'collectives'. These are stores where a number of different sellers rent stalls, so there is a wide range of items and subjects to shop for. If you are shopping in collectible stores or collectives, you will find more and more plastic kitchen pieces appearing as dealers and sellers become aware of older plastic's value. Don't be surprised if some shop owners mislabel a plastic item, calling any old plastic Celluloid or Bakelite. This is probably out of misinformation, and is another good reason to learn to identify your plastics. However, if you like a piece and know what it is, it really doesn't matter. The more desirable kitchen items (canisters, salt and peppers, tumblers, flatware) are being seen in these establishments and seem to be selling well. You will be paying higher prices for these items so examine them carefully. They should be in excellent to very good condition. Remember, *buy now!*

A few final notes, hints and observations:

1) I have tried to show as many examples of plastic kitchenware and dinnerware as possible in this book. There are hundreds more pieces to be discovered, and they are turning up every day. Time, money, and deadlines have been my only limitations.

2) Some pieces that are unmarked may be newer pieces. Especially when shopping thrift stores and garage sales, you're on your own. Many enthusiastic friends have brought me 'neat' plastic kitchen pieces they have found which have turned out to be new. I have tried to use my judgement and when possible indicate that a piece *might* be newer. More often than not, if I am in doubt about a piece's age, I have chosen not to show it in this book. Any input from you, the collector, would be helpful.

3) You will find pieces that are marked Made in Hong Kong or just Hong Kong. These pieces are worth collecting for two reasons. First, plastic items from Hong Kong have been imported into this country since the 1950s (the absolute newest imported kitchenware items are from Taiwan, China or elsewhere and are marked accordingly). Second, with the relinquishing of British rule over Hong Kong in 1997, items marked "Made in Hong Kong" will soon become collectible. (For that matter keep an eye out for pieces marked "West Germany," though I'm not sure how much plastic kitchenware they imported.)

4) Some of the items listed in this book might not seem to qualify as kitchen items per se. You need to understand the changing concept of the kitchen during the post-war years. In the 1930s (and up through the war years) the kitchen was seen as the 'Kitchen Laboratory' — chrome-edged streamlined design, white steel cabinets and appliances, and technical lighting. The efficiency experts counted each step to get from the refrigerator to the sink and from the sink to the range and so on. In these minimalist and sterile environments, isolated from the other rooms of the house, Scientist Mom created her culinary feats with hopefully (in the efficiency experts' eyes) a minimum of effort. By the post-war years everyone (especially housewives) screamed "Enough!!" Sure, they wanted all the elements of efficiency, but they also wanted their kitchens to be livable and charming as well. Color and decoration swept into the kitchen and swept away the cold all-white look that had dominated the kitchen environment for so long. (To be honest, color had been slowly entering the kitchen since the 1930s, but that's a whole book in itself). Equally important was the concept of making the kitchen more livable by making it more multi-purpose. *How To Improve Your Home For Better Living*, published in 1955, suggests the following:

"The work center can be supplemented with other related activities that turn the kitchen into a multi-purpose room. Mother's work can be simplified greatly by such an arrangement. For example, the laundry can be brought up from the cellar and the sewing machine moved down from the attic. A desk, where bills can be checked and records filed, can serve as the nerve center for household administration. This should have a telephone, shelves for cookbooks and recipes. If there are young children in the family, it is advisable to have a small play area nearby where the boys and girls will be under mother's watchful eye. Eating in the kitchen also simplifies mother's work."

The kitchens we are focusing on were used for a number of purposes. So, don't be surprised to see flyswatters, wastebaskets, and cleaning supplies listed in this book.

Inviting, practical service for festive or special occasions. Barbecue Set in newest Styron-tone colors complements any setting.

Wise hostesses save steps with this generously proportioned pitcher of sturdy, acid-resistant Styron (Dow polystyrene). . . . Iced drinks greet guests handsomely in these pint size tumblers of easy-to-handle, easy-to-clean Styron.

the smart hostess 'follows suit' with

bright..smooth easy to clean

Plastics Housewares

↓

MADE OF STYRON®
A DOW PLASTIC

look for the **STYRON** label

It's your assurance that the product has been evaluated for proper application of plastics, sound design and good workmanship.

Quick serving tricks are yours with this Susan Set of colorful Styron! Removable compartments serve salads, desserts, snacks anytime.

Any salad makes a beautiful entrance in this graceful bowl of smooth, bright Styron.

Bread is always ready for a party in this tight-closing, easy-to-clean box that's decorative, too, in Styron.

The smart hostess has an extra trick up her sleeve these days! She plans her parties with gay, work-saving plastics housewares made of Styron (Dow polystyrene). She knows there's all around utility in Styron housewares . . . from crystal clear refrigerator boxes that keep her food fresher, to colorful serving dishes that add a festive air to the simplest party snack. And Styron housewares are so easy to handle, easy to clean, they cut party work to a minimum. You'll find, too, that the Styron label means housewares that are as practical as they are decorative.

THE DOW CHEMICAL COMPANY
Plastics Department—PL-136 • *Midland, Michigan*

look for STYRON also in . . .

Coat Hangers

Refrigerators

Wall Tile

DOW
plastics

Entertaining with Styron gifts made by Dow Chemical Company.
From *Good Housekeeping Magazine*, 1952.

A hot plate with a metal plate and black plastic casing, marked "An All States Wire and Metal Product," and an electric coffee pot in avocado and black plastic, marked "Regal Poly Perk by Regalware." This coffee pot is hard to find complete with its insides, cord, and lid. Also pictured are a coffee canister in metallic Polystyrene by Republic, a metal and ceramic trivet, and a metal coffee scoop for Caswell's Coffee.

Chapter Three
Kitchenware Items

Kitchenwares are defined as any items associated with the five basic activities in a kitchen: storage, food preparation, cooking, serving and clean-up. I have expanded the definition to include items for serving in the dining room and other accessories for social functions which involve serving food or beverages. The items covered in this book are entirely plastic, except for certain appliances which are partly plastic or encased in plastic housing (the body of the appliance), flatware and utensils with plastic handles, and items like dispensers which are combinations of metal or glass and plastic.

Research into the background and histories of kitchenware plastic companies has turned up very little, since most companies are non-existent or extremely hard to locate at this time. I have tried to provide as much information as possible. The following categories of plastic kitchenware items are listed alphabetically. All items are from my own collection unless otherwise noted.

APPLIANCES

This vast field could comprise a book of its own. Only a small handful of examples is presented here, just to give you an idea. Almost all the plastics mentioned in the first section of this book were used in some way with appliances. I define appliances as large items, usually electrical or mechanical, which are made of a combination of plastic and other materials, usually metal and/or glass. Smaller appliances I have listed under "Gadgets," as they are usually non-electrical and all plastic.

Ice-crushers and hotplates were particularly popular in the 1950s and you will find many examples. Pink and turquoise appliances were big in the mid-1950s. In the 1960s General Electric introduced a whole line of avocado-colored kitchenware appliances, thus starting the avocado color craze.

Presto automatic hot dog cooker in red and white plastic; Westinghouse automatic electric can opener in white and turquoise plastic. Notice how the word 'automatic' keeps popping up.

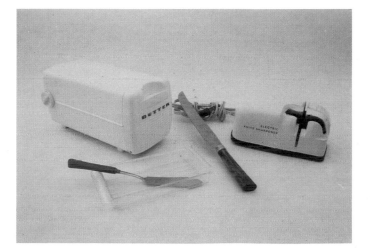

(Left to right) A butter heater, white with removable clear dish; a knife sharpener, white Urea, marked "John Oster Mfg. Company." Also pictured are a butter knife with a green Catalin handle and a table knife with a marbelized Catalin handle. *Butter heater courtesy of Modern Times, Portland, Oregon*

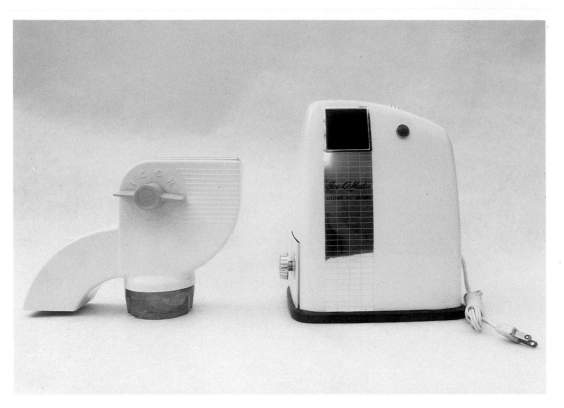

(Left to right) An icer attachment for the Osterizer, white with an aqua handle; an Ice-O-Matic electric ice crusher with a white plastic body, marked "Rival Mfg. Company Kansas City, Mo." *Courtesy of Palookaville, Portland, Oregon.*

BARWARE

The repeal of Prohibition in 1933 revived many industries involved with alcohol and the serving of alcohol. Distilleries as well as bars and taverns proliferated. Home entertaining involving alcohol returned, as did cocktail cabinets and home bars. Glassware, china, and metal companies started producing bar and drinking items again. These items are very popular with Depression glass and Art Deco enthusiasts. Some of these items, including cocktail shakers, were made with plastic components. Serving drinks also fit in well with the post-war concept of 'relaxed entertaining'. Plastic ice buckets, shakers, coasters and 'swizzle' sticks were made in large quantities to fill this need. Wall mounted ice crushers with plastic bodies can also be found. Many plastic tumblers, glasses, and trays fit nicely into this category.

Catalin-handled drink stirrers and pickle forks. *Courtesy of Cat Yronwode.*

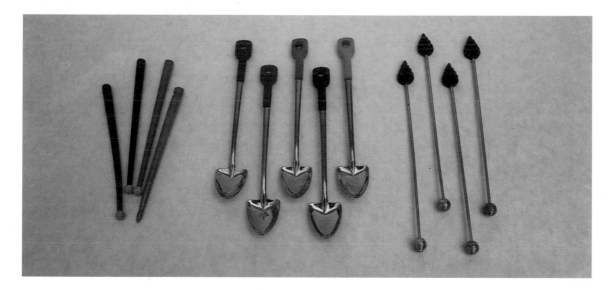

Catalin drink stirrers in maroon, green, orange, and butterscotch; drink shovels marked "N.S. Company Durever"; black Catalin leaf-handled stirrers. *Courtesy of Cat Yronwode.*

Catalin barware. (Left to right) Butterscotch and red holder for straws, etc., marked "Rheingold Lager Beer"; butterscotch shotglass caddy. *Courtesy of Cat Yronwode.*

Catalin barware: a butterscotch and black match holder (candle unscrews); a black semi-circle matchbook holder; bottle stoppers in shapes of a scotty dog, a bird, a rose, and a terrier. The rose has the cork still attached. *Courtesy of Cat Yronwode.*

(Left to right) An ice bucket made of clear acrylic with gold speckling and white insert (lid missing); an ice bucket in brown and tan Polystyrene, marked "Lustro-Ware."

A naturalistic leaf coaster and stirrer set with rubber leaves and plastic 'twig' stirrers; Ubangi cocktail mixers, marked "Made in Hong Kong 1951"; a bridge coaster and stirrer set in blue and red with a clear plastic case.

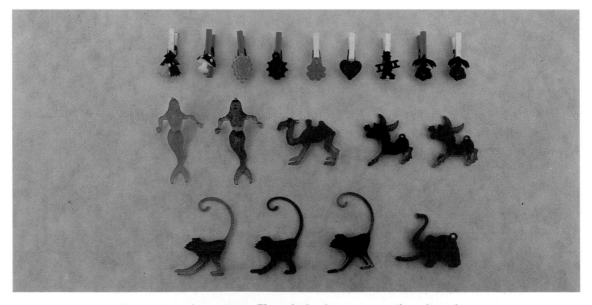

Plastic glass decorations. These little objects were either clipped or hung on the rim of cocktail glasses for "humorous adornment." The top row are all clips.

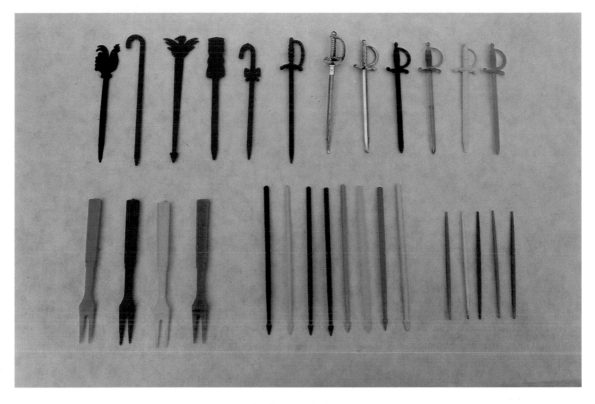

Cocktail and olive picks.

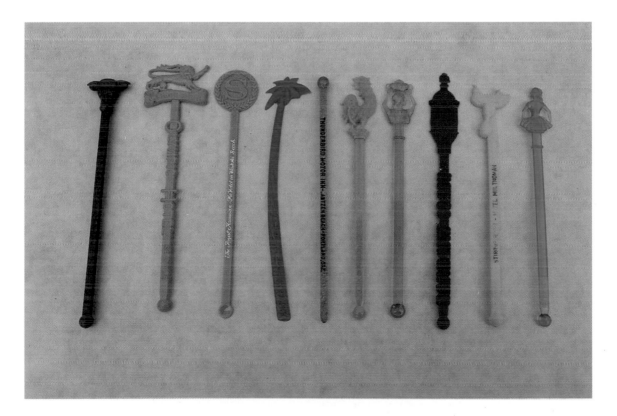

Various cocktail stirrers, or 'swizzle' sticks.

(Left to right) A battery-powered shaker, marked "Made in Hong Kong," in red and clear Polystyrene; Sweetheart "Sparkles" transparent plastic straws; a straw holder in yellow and clear Polystyrene. At front is a cocktail strainer with a marblized Catalin handle.

A collection of coasters. (Top row, left to right) Turquoise and yellow Polystyrene (whistle missing); zodiac Polyethylene; black Polystyrene marked "Styson Art Prods. NYC"; orange and yellow turtle Polystyrene. (Bottom row, left to right) Yellow and blue marked "Rogers Prod."; pink Polyethylene; mottled green and white Polystyrene.

More coasters. (Top row, left to right) Green, yellow, red, and blue set in Polystyrene. (Bottom row) Leaf-shaped styrene coasters by Gothamware in chartreuse, salmon, gray, jungle green, and red.

(Top row) Owl coasters in styrene. (Bottom row) Transparent styrene coasters marked "Beacon Products."

(Left to right) La Stella coaster set in pink styrene with holder, marked "Poly Plastics, Leominster, Mass"; black Melamine coaster/ashtray combination, unmarked; Nautical motif styrene coaster set and holder in black with gold, marked "PM Coaster Corp. Sarasota, Fla."

(Left to right) A chrome shaker with butterscotch Catalin handle and finial, 12" tall; a chrome pitcher with butterscotch handle marked "Chase," 9" tall; chrome shaker with green Catalin handle and finial, 12" tall. *Courtesy of Cat Yronwode.*

BOWLS

Listed here are mixing bowls and storage bowls. The majority of them are made of Polyethylene as this has proven to be the most adaptable and functional plastic for the rigors of cooking. Look for the names Plas-Tex and Beacon, two companies with a long history of producing utilitarian kitchen items. You will be surprised at how many types and forms the simple bowl shape can come in. Mixing bowl sets usually came nested (fitting inside each other) in groups of four to six. Storage bowls usually came with snap on lids (many are missing). Many bowls took a beating (literally) and you will be hard pressed to find one in excellent shape — scratch marks and dulling being the major damage. Serving bowls are usually larger and are listed under "Serving Pieces."

Promotional photo showing unbreakable mixing bowls of polyethylene. Pictured are Blisscraft of Hollywood bowls. From *Modern Plastics Magazine*, 1955. *Photograph courtesy of the Bakelite Company.*

IDEALWARE
MIXING BOWLS

An advertisement for mixing bowls by the Celanese Corporation. From a 1952 *Good Housekeeping Magazine.*

This Melmac mixing bowl is durable, molded in a colorful spatter pattern by Plastics Manufacturing Co., Dallas, TX. From *Modern Plastics Magazine*.

The four-piece Serv Mix n' Stor Bowl set in yellow, salmon, turquoise, and mustard yellow Polyethylene. Each bowl came with a lid.

Three yellow Polyethylene bowls. (Left to right) Swoop-shape, marked "Plas-tex," and marked "Raymond Loewy Associates."

(Left to right) Polyethylene bowls made by Beautyflex in Montabello, CA. The center back piece is a colander.

Polyethylene bowls. (Left to right) With swirled ribbing, marked "Shamrock Neatway"; in center, with horizontal ribbing and no mark.

(Left to right) Pink Polyethylene bowls (10" and 6" diam.) with lids marked Stanley-Flex, and a red and white Polystyrene bowl.

(Left to right) An institutional green bowl, 9.5", marked "Boonton"; a multi-colored Melamine bowl, 10", marked "Texasware"; a turquoise bowl, 9.5", marked "Apolloware." *Texasware courtesy of Darryl Ballini.*

(Left to right) Mixing bowls in tan multi-colored and green multi-colored designs. *Courtesy of Old Town Antique Market, Portland, Oregon.*

BUTTER DISHES

Pre-war glass butter dishes were larger and made to accommodate the one-pound butter slabs that were sold in those days. These might look funny to you now. Larger glass dishes were also made to store butter in the refrigerator. These are sometimes called 'Refrigerator' or 'Leftover' butters. Quarter-pound sticks of butter (four to a pack) became the standard by the 1950s. Most plastic butter dishes of the 1950s will be the size to fit the sticks. Besides the butter dishes that are part of Melamine dinnerware sets, you will find numerous independent dishes. Clear lids were popular in Polystyrene items. Most later examples are of Polyethylene. There is no such thing as an open butter dish — only a bottom estranged from its top.

Butter Dishes. (Top row, left to right) Corn-shaped in green/yellow Polyethylene; in yellow Polystyrene. (Bottom row, left to right) Yellow Polystyrene; yellow and frosted Polyethylene marked "Lustro-Ware."

Butter Dishes. (Top row, left to right) Green/clear Polystyrene; yellow/frosted Polyethylene marked "Blisscraft of Hollywood." (Bottom row, left to right) Charcoal and pink Polystyrene marked "Burrite"; Beige/frosted Polyethylene.

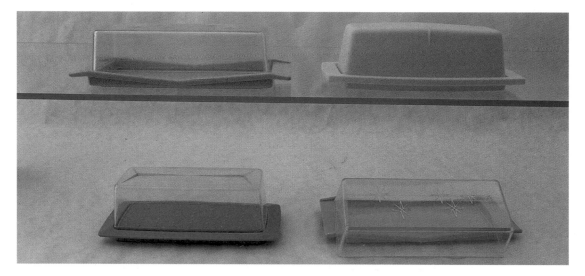

Butter Dishes. (Top row, left to right) Pink/clear Polystyrene; ivory Polyethylene marked "Max Klein Inc." (Bottom row, left to right) Turquoise/clear Polystyrene marked "Deka Elizabeth, N.J."; Turquoise/clear Polystyrene with gold starbursts on lid.

Butter Dishes. (Top row, left to right) Pink Melamine swoop-shape; Gray/pink Polystyrene (handle on top missing) marked Plasmetl. (Bottom row, left to right) Copper-colored Polystyrene (lid missing) marked "Blisscraft of Hollywood"; yellow Melamine.

Canisters and Cookie Jars

Webster's Dictionary defines a canister as "a useful thin metal container used for holding dry foods." Except for the somewhat biased (and humorous) insistence on the word "metal," this is a pretty accurate description. What is the difference between a canister and a jar? Webster's dictionary again: "A jar is a cylindrical glass or earthenware vessel with a wide mouth and usually having no handles." Of course, today jars and canisters can be any shape, and may be made from any material — even plastic!

Plastic canisters are among the most popular and recognizable items in this field of collecting. Individually or in a whole set, they look great sitting on a kitchen counter or shelf. Particularly desirable are the sets made by Columbus Plastic Products as part of their Lustro-Ware line, followed by The Burroughs Manufacturing Co.'s Burrite line. Nearly *everyone* who sees one of these canisters says "OH, I remember those!" Many other companies made canisters as well, so don't pass these by. Canisters were sold in sets of four — Flour, Sugar, Coffee and Tea, from largest to smallest. Look for screened-on lettering in good condition. Missing lids defeat the purpose of a canister, but buy it if the item is in good shape with good lettering. Your chances of finding the matching lid are still good, or you might be able to match a bottom with sharp lettering to a lid that came with a canister that suffers from worn lettering.

Plastic cookie jars are listed here, as they are actually canisters by definition.

An advertisement for the Lustro-Ware Kitchen Ensemble. An economic bridal gift — an 8 piece set for just under $10.00. One of their most successful items. From *Good Housekeeping Magazine,* 1953.

Polystyrene canisters. (Left to right) Red and white coffee, marked "Burrite"; red with white printing, marked "Popeil Brothers"; red and white coffee and sugar, marked "Nupco." Last two canisters *Courtesy of Old Town Antique Market, Portland, Oregon.*

Polystyrene canisters. (Left to right) Yellow and white sugar and tea size with 'see-through' windows, marked "Nudell"; pink and yellow with clear tops. Last two canisters *Courtesy of Palookaville, Portland, Oregon.*

A four-piece Polystyrene canister set in yellow with white lids, made by Loma Plastics. *Courtesy of Palookaville, Portland, Oregon.*

Polyethylene canisters. (Left to right) Yellow flour; blue coffee and tea with white lids; mustard yellow tea canister with white lid.

Canisters. (Left to right) Tea, turquoise with black lid, and coffee, yellow with white lid, both from the same Polyethylene line; a sugar-size Polystyrene canister, turquoise with white lid. All have gold decoration.

A four-piece Polystyrene set of canisters, in yellow with white lids and tropical decals, marked "Burrite."

Polystyrene canisters. (Left to right) Tea, with aqua body and knob and a white lid; coffee, with pink body and knob and a charcoal lid; sugar, with yellow body and charcoal knob.

A four-piece yellow Polyethylene set of canisters marked "Plas-Tex Corp."

Canisters. (Left to right) Sugar and coffee in turquoise; coffee and tea in yellow with white lids, marked "Burrite."

Cookie jars. (Left to right) Pink with charcoal lid and pink knob; red with white lid, marked "Columbus Plastic Prods." Also pictured is a metal tidbit tray in turquoise, white, and gold.

CHILDREN'S ITEMS

Even in shops, very few children's feeding items can be found, despite the incredible quantities made through the 1940s, 1950s, and 1960s, not only for retail sale but also as premiums and promotions. Plastics were an absolute boon to all areas of the children's industry — playthings, utilitarian objects, and, of course, feeding items and dishes. The 'childproof' aspect of plastic dishes and feeding sets was one of their strongest selling points. Many a toddler sat for hours banging his or her feeding dish or plastic cup in an attempt to test its durability. Well, maybe that explains why not much of it has survived. You will find pieces made not only of Melamine but of Urea and Polystyrene as well. Divided infant dishes (the ones with the hollow compartments underneath for holding hot water to keep baby's food warm) are highly desirable, but make sure the stopper is not missing. Have you checked the prices of older ceramic ones in shops? WOW!! The most desirable children's items have recognizable celebrities or cartoon characters on them. You will vie with collectors of these two popular categories for items of interest, and depending on the character or image, expect to pay slightly higher prices than you would for a plain dish, bowl, or tumbler.

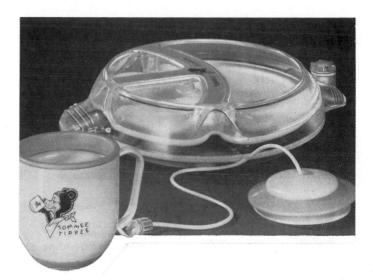

An advertisement for Koppers Plastic Company showing the Tommee Tippee three-piece set in new, improved Polystyrene. From *Modern Plastics Magazine*, 1957.

"Baby's Dyner" feeding set. Molded of Urea by Auburn Button Works. From *Modern Plastics Magazine*, 1939.

A charming children's tea set in Urea with Alice of *Alice in Wonderland* on the side. From *Modern Plastics Magazine*, 1937.

Children's flatware pieces with Catalin handles. (Left to right) A green-handled spoon; a red-handled fork; a red-handled spoon. *Courtesy of Cat Yronwode.*

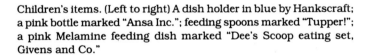

Children's items. (Left to right) A dish holder in blue by Hankscraft; a pink bottle marked "Ansa Inc."; feeding spoons marked "Tupper!"; a pink Melamine feeding dish marked "Dee's Scoop eating set, Givens and Co."

Children's items. (Back row, left to right) A divided baby dish metal with white body and handles; a yellow Polystyrene cup boot-shaped; a 3" white cup marked "F&F Mold and Die"; a blu straw with a cat on the end; a white and yellow Polyethylene bab dish marked "Westland Plastics Inc." (Middle row, left to right) 7" plate with a clown, made by International Molded Plastics; a 4 white bowl with Mary Poppins, marked "Sun Valley"; a 7" plat with Hanna Barbera characters, marked "Boontonware". (From row, left to right) Flatware, including a red cake knife, a yello butter knife, and place settings in yellow, turquoise, and pin marked "Gothamware."

Children's items. (Left to right) An eight-piece Polyethylene popsicle-making kit; a Polystyrene Home Treat Ice Pop kit. *Polyethylene set courtesy of Modern Times, Portland, Oregon*

Children's items. (Left to right) A pink mug with a clown, marked "F&F Mold and Die"; a blue Captain Kangaroo cup with 'googly' eyes; a red 'silly strawberry' mug marked "Pillsbury Corp."; light blue Porky Pig mug with 'googly' eyes, marked "Eagle." *Courtesy of Modern Times, Portland, Oregon.*

CLOCKS

Kitchen clocks have a look of their own. They are usually a different size (approximately 7" x 7") than other household clocks and are wall-mounted. They always typify the colors and styles popular with kitchens at the time they were made. Kitchen clocks 'went plastic' probably around the late 1930s. Make sure a clock is working before you buy it. Also, check for cracks. Figural clocks are harder to find than the plain ones.

EVEN ELVIS DOESN'T HAVE A CREED

The famous Kit-Cat Klock is one of the most recognizable icons of popular culture. This famous "kitsch" item was created in California to brighten homes during the mid-1930s Depression. It must have worked, because Kit-Cat Klocks have maintained their popularity up to this day. Every three minutes for the last 50 years someone has purchased a Kit-Cat Klock, says Woody Young, head of the California Clock Company.

As popular as the original clocks are, Kit-Cat doesn't just sit on the wall with his rolling eyes, big smile, and wagging tail. There are now official Kit-Cat shirts, towels, postcards, greeting cards, and other items. There's an international Kit-Cat Fan club with over 2,000 members, who adopt Kit-Cat's positive philosophy towards life. They live out the Kit-Cat Creed: "Put a smile on everyone's face; Love in everyone's heart; Energy in everyone's body; and Be a positive force in everyone's life!"

Kit-Cat Klocks can be purchased today in the original colors at clock shops or gift stores. They are also available through the California Clock Company. Write to them at P.O. Box 827, San Juan Capistrano, CA 92675 for information and an order form. They also sell KC Bear, Fifi, the French Poodle, and Professor Timebelly (the Owl). As of 1995 the company changed from electrical cords to battery-operated power for their cats, poodles, bears, and owls.

In Wall Clocks...it's Westclox!

MELODY Electric Wall Clock is adaptable to any room, any color scheme. Smartly-styled, Melody mounts flush on wall; excess cord is neatly concealed. Easily removable case ring comes in a wide variety of colors. $6.95.

An advertisement for Westclox (the makers of Big Ben?). Shown here is a styrene model called "Melody." From *Good Housekeeping Magazine*, 1952.

Two clocks by Westclox: a pink body with a black face, and a pink and black body with a white face. *Black-faced clock courtesy of Old Town Antique Market, Portland, Oregon*

38

Two clocks by General Electric: a red body with a white face, and a peach body with a black face.

Two clocks: a brown body shaped like a bread board with a white face, by General Electric, and a yellow daisy with a white face, by Timex.

CONTAINERS

This is a catch-all category and covers open and closed containers and keepers. These containers could be used for storing or holding food, gadgets, or utensils. Some containers you discover will actually be canisters with the lids missing. See the "Refrigerator Dishes" section for more examples of what might be called containers.

Vegetables "Bloom" in my Tri-State Permacrisp!

I shop for "greens" but once a week, and how fresh my Tri-State rigid plastic refrigerator "garden" keeps them! The self-seal Permacrisp lid locks in the "dew"...yet a fingertip on the pull-up tabs, and it's open! The ingenious ridged bottom keeps cold air circulating on all surfaces, makes for easy stacking. And the chip-proof, feather-light plastic, with the rounded corners washes crystal-clear! $1.79

Look for the Tri-State Permacrisp with my Trudy Star label at your hardware, housewares stores, supermarket, 5 & 10, or department store.

FREE! "Keeping House with Plastics." Household hints and amusing ideas—galore! Write: Trudy Star, c/o Tri-State, Henderson, Ky.

Trudy Star says: my Tri-State
TWIN BREAD BOX
Keeps Bread Fresh for a Full Week!

Clean,
Crystal-Clear
RIGID PLASTIC $2.69

At leading hardware and dept. stores
or write...Trudy Star, c/o Tri-State, Henderson, K

An advertisement for the Tri-State Permacrisp vegetable crisper and for the Tri-State Twin bread box. Both are endorsed by Trudy Star, Tri-State's 'happy homemaker' whose "Trudy Star Label" appeared on Tri-State's products, apparently assuring quality. From *Good Housekeeping Magazine*, 1954 and 1955.

An advertisement for Poly-Flex food containers and mixing bowls by Republic Molding Co. of Chicago, IL. From *Good Housekeeping Magazine*, 1954.

A white and clear Polystyrene Lustro-Ware cake keeper with roosters, and a brown and clear Polystyrene pie keeper, patented in 1954. *Cake keeper courtesy of Modern Times, Portland, Oregon;*

Red and clear and yellow Polystyrene cake keepers, both marked "Lustro-Ware."

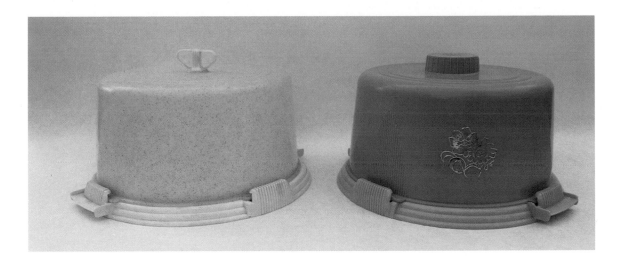

Two containers, both marked "Look Lift Cake Cover, Trans Spec Company." One has a white base and a frosted gold-speckled lid; the other has a beige base with a mustard yellow lid. The lids are Polyethylene and the bases are Polystyrene.

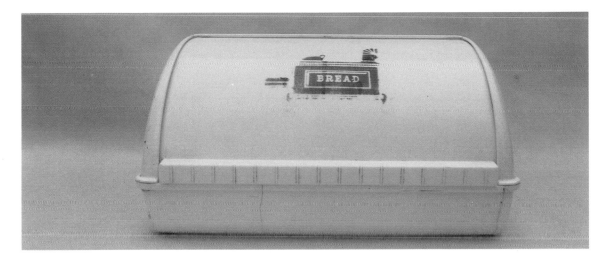

A white Polystyrene bread keeper, marked "Burrite by Burroughs Mfg. Co."

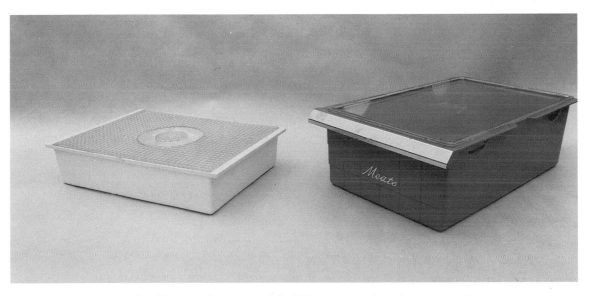

A refrigerator keeper in white Polystyrene with a clear, textured top, and a salmon-colored meat keeper for the refrigerator with a clear top. *Meat keeper courtesy of Palookaville, Portland, Oregon.*

A clear water jug for the refrigerator, and a red and clear Polystyrene bread keeper marked "APCO."

An apple-shaped covered jar with gold speckles, marked "VV's," and a clear humidor-type container with desiccant crystals in the knob.

An open container in dark brown Bakelite, marked "Continental Plastics of Omaha"; a water or juice jug in aqua and white Polyethylene, marked "Plas-tex Corp."; an open container in red Polystyrene, with rubber feet.

FLATWARE

Also called silverware or cutlery. Most of the pieces shown here are metal utensils with plastic handles. The vast array of pre-war pieces are from the collection of Cat Yronwode. The handles are made of Catalin and were extremely popular between the mid-1930s until shortly after the war. (The photographs of Catalin flatware shown here are grouped by shape of handle.) You can still find Catalin-handled cutlery, but the supply of samples in good condition is dwindling. This is another 'kitchen collectible' you will find in shops. Later period plastic handles were made from other plastics. In 1953, Melamine was first used on cutlery handles and became used extensively for that purpose. By the mid-1950s flatware handles made of Zytel, a nylon resin, were introduced. Some pieces of flatware are wood handled with plastic laminate. All-plastic cutlery was usually cheaply made for functions like children's birthday parties. (Remember the fork tines breaking easily as you dug too hard into that slice of birthday cake?) Older, better-quality pieces were made to go with picnic sets.

Flatware with Catalin handles, in two-tone butterscotch and red bridge pieces.

Flatware with Catalin handles in various two-tone combinations of butterscotch and black.

43

Flatware with Catalin handles, in two-tone butterscotch and translucent.

Flatware with Catalin handles. (Left to right) Three in solid colors; three marbelized; one pearlized.

Flatware with Catalin handles and 'bulbous' bottoms.

Flatware with Catalin handles and 'chevron-like' bottoms.

Flatware with Catalin handles and 'chevron-like' bottoms.

Flatware with Catalin handles, some with 'rounded' bottoms and
others with 'chevron-like' bottoms.

Flatware with Catalin handles and Celluloid inlay.

Flatware with Catalin handles and Celluloid inlay.

Flatware with Catalin handles and 'flat' bottoms.

Knife and fork flatware sets with Catalin handles in ivory, redwood, and butterscotch.

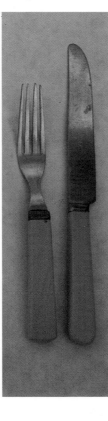

Flatware with Catalin handles and 'curved' bottoms.

48

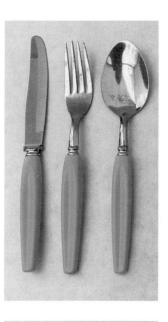

Flatware with Catalin handles and 'sweep-shape' bottoms.

Miscellaneous flatware with Catalin handles.

Flatware place settings, all marked "Stainless," with turquoise, brick red, and ribbed pink handles.

49

A flatware place setting in turquoise Polystyrene, and knives marked "Sheffield" with ivory-colored handles.

GADGETS: FOOD AND NON-FOOD

The 'gadget' category is all-encompassing, covering just about everything that hasn't been covered in the other categories. It can be broken down to two different subdivisions: 1) gadgets used in the preparation of food and 2) gadgets that were used in the kitchen but not directly for food preparation. A gadget is defined in Webster's dictionary as: "a small specialized mechanical device." I have taken liberties in including non-mechanical devices as well. This is a really fun category to collect and assemble. Many smaller gadgets can still be found in their original boxes or wrappers.

(Left to right) Chartreuse Cellulose Acetate recipe file; yellow translucent Polystyrene recipe file, marked "Eldon Californiaware"; a "Cook Index" with an ivory-colored body and a silver top; two yellow Polystyrene clover-shaped nut dishes. Also pictured: cooking clips recipe book, and a blue tile trivet with a decal.

(Left to right) Flyswatters from Lyle's Thrifty Food Market, Portland, OR; Shop-Rite Market, OR; Oregon Army National Guard; and Clark and Moore, IN.

More flyswatters. (Left to right) Lustro-Ware, Plasti-Swat, and one in avocado and white.

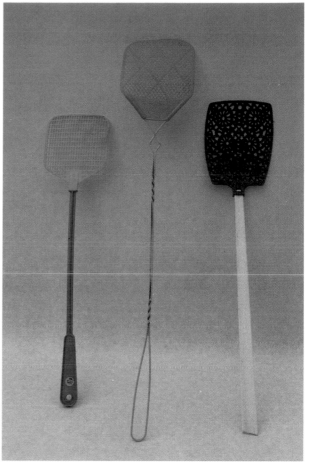

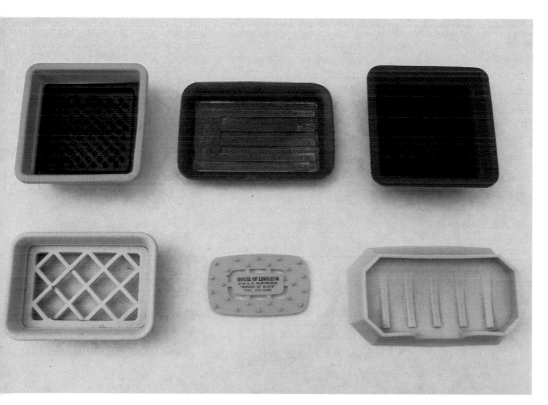

Soap dishes. (Top row, left to right) Marked "Lustro-Ware"; marked "Daisy Soap Tray"; marked "Lustro-Ware." (Bottom row, left to right) Pink Lustro-Ware; turquoise "House of Linoleum"; pink holder. All are made of Polyethylene except the "Daisy" dish, which is rubber.

(Top row, left to right) A chef spoon rest in white; tea bag holders in green and yellow; and yellow lovebirds spoon rest, marked "Admiration." (Bottom row, left to right) A red pear spoon rest, marked "Fuller"; a white potholder hanger shaped like a teapot; a red fish spoon rest, marked "Plastica, North Wales, Pa." All pieces are Polystyrene.

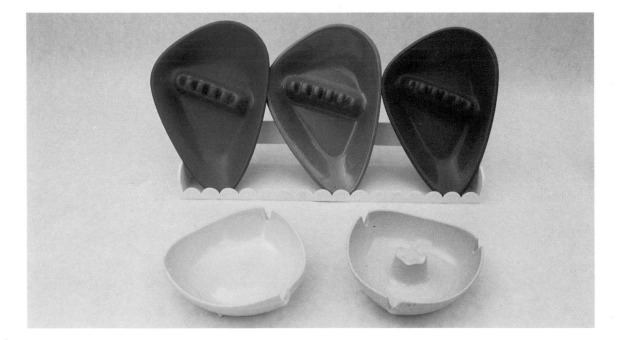

Ashtrays. (Top row, left to right) Dark green, brown, and orange ashtrays marked "Anholt Ashtray." (Bottom row, left to right) Yellow and institutional green ashtrays, one with a 'snubber' in the middle. The Anholt ashtray's patent number on the back dates it to 1943, but these ashtrays are still being made today.

Ashtrays. (Left to right) Yellow with orange mottling, marked "Irwin-Willert Co."; white with gray mottling, marked "Deep Dish Ashtray"; brown with white mottling, blob-shaped. The Irwin-Willert company is still in business.

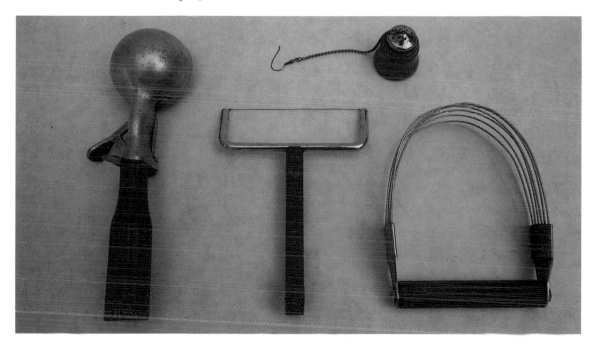

A red Catalin-handled scoop; a red Catalin tea ball; a red Catalin-handled cheese slicer; a green Catalin-handled pastry blender, marked "Androck." *Courtesy of Cat Yronwode.*

(Left to right) A green and yellow Catalin butter pat maker with a flower-shaped mold in the bottom (depress the handle onto butter, release the handle, move to a plate and depress again to make a small round pat of butter with a flower imprint); a wood and metal cheese slicer with red Catalin handles and a butterscotch scotty on the slicer. *Courtesy of Cat Yronwode.*

53

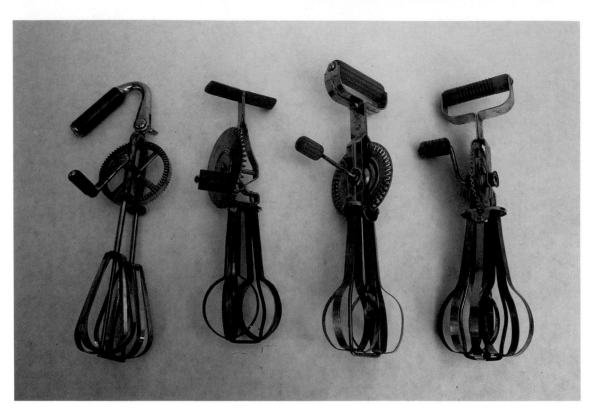

Catalin-handled eggbeaters. The item on the left is marked "Roos Mfg. Co."; the rest are marked "Androck." *Courtesy of Cat Yronwode.*

Catalin gadgets. (Left to right) A red crinkle cutter; a red parsley chopper; a red peeler marked "Englishtown"; a green peeler marked "Androck"; a red peeler marked "T & S." *Courtesy of Cat Yronwode.*

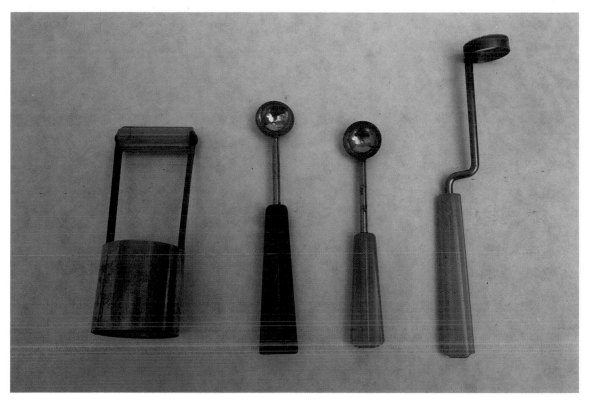

Catalin gadgets. (Left to right) A butterscotch-colored corer; a black
melon ball scoop; a green melon ball scoop; a butterscotch corer.
Courtesy of Cat Yronwode.

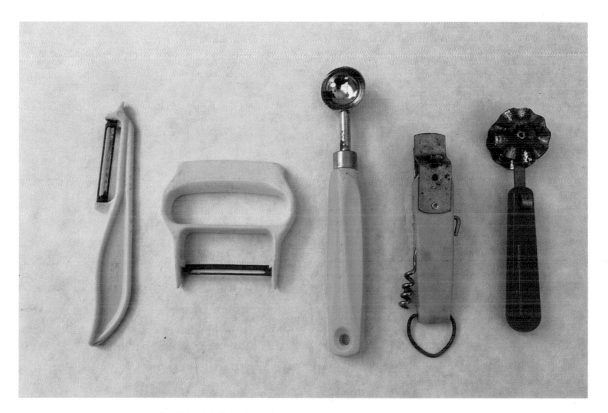

(Left to right) An ivory-colored peeler; a yellow peeler; a turquoise-
handled melon ball scoop; a yellow can opener; a red-handled
dough wheel.

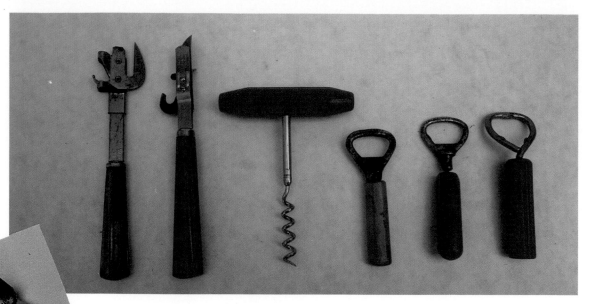

Catalin gadgets. (Left to right) Red and green can openers; a red corkscrew; amber and red bottle openers. *Courtesy of Cat Yronwode.*

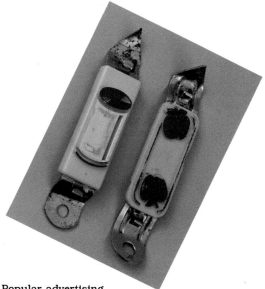

Can and bottle openers with plastic bodies. Popular advertising gimmicks.

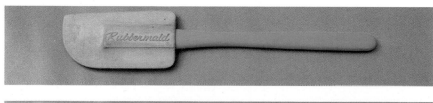

(Top to bottom) Plastic-handled rubber scrapers in red and blue; the blue is marked "Rubbermaid." A baster with a plastic tube and a pink rubber squeezer.

(Left to right) A yellow pastry brush with a cartoon chef (may be newer); a pink comet-shaped pastry brush and wheel; a turquoise pastry brush advertising "Herb the Hermit"; a red and white all-purpose cutting wheel; a light green melon ball scoop. All items are Polyethylene.

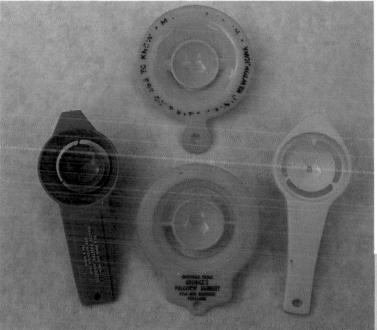

Polystyrene egg separators in turquoise, yellow, and lemon yellow. Egg separators were another popular premium for advertising. Many make silly references to eggs, such as "A good egg to know - George's Market."

(Top to bottom) A rolling pin with a yellow plastic body and yellow and red handles; a rolling pin with white plastic handles and salmon-colored metal body.

MEASURING DEVICES

Measuring devices generally fall into two categories — measuring cups and measuring spoons. Both came in sets of four: cups in 1 cup, 1/2 cup, 1/3 cup and 1/4 cup; and spoons in 1 tablespoon, 1 teaspoon, 1/2 teaspoon and 1/4 teaspoon. There are also large individual measuring cups with increments marked on the side. I was surprised at the vast number of examples I was able to acquire. All plastics are represented here with Polyethylene (of course) leading the pack. Red is a common color.

The "Wet'n'Dry" measuring cup. Manufactured in Polyethylene by Westland Plastics Inc., Los Angeles, CA.

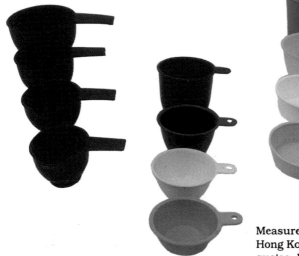

Measures. (Left to right) A red Polystyrene set, marked "Made in Hong Kong"; a red, blue, white, and yellow Polystyrene set; a turquoise, lemon, white, and pink Polyethylene set, marked "BW."

A variety of Polyethylene measures. (Left to right) A two-sided cup; a two-cup measure by Plas-tex; a high lipped cup, marked "Merrymaid Plastics, Cleveland, Ohio."

Various measuring cups. The green cup (top middle) is made of Urea and is marked "Safetyware."

TUMBLERS AND MUGS

A tumbler is a straight-sided drinking glass with no handles. It can be small like a juice glass or almost 8" tall. The number of examples is amazing, as is the array of colors. Examples in both Polystyrene and Polyethylene are easily found. You will need to use your judgement to discern the old from the new. Many plastic companies made tumblers, though some are unmarked.

The category of mugs seems endless. A mug, of course, is handled and generally squatter than a tumbler (but not always). All sizes and styles are represented here. Inexpensive Polystyrene ones are still being made today. Look for backstamps! Printed advertising on the sides is usually of recent vintage, but that could be a whole field of future collectibles.

Polystyrene mugs and tumblers. (Top row, left to right) An orange tumbler, 5" tall; a yellow and white mug, 4" tall, marked "Deka Elizabeth, N.J."; a Lifesavers mug, 3.25" tall, marked "Deka"; a beige tumbler with a blue daisy, 5" tall, marked "Designed by David Douglas. Made of genuine Accalac." (Bottom row, left to right) A translucent pink tumbler, 3" tall, marked "Avalon Products"; a red tumbler, 3" tall, marked "Allied Plastics USA"; two mugs, 2" tall, marked "Acrylite ware. National Plastics Company for TWA."

Polystyrene mugs and tumblers. (Top row, left to right) Mottled tumblers in yellow, red, and blue, 3.75" tall, and multicolored, 4.50" tall; speckled mugs in sea green, salmon, pink, and yellow, 3" tall.

Polystyrene mugs and tumblers. (Top row, left to right) Aqua mugs, 2" and 3" tall, marked "Shamrock Neatway"; two ivory-colored tumblers, 4" tall, hand painted but badly peeling, possibly Urea. (Bottom row, left to right) A mustard yellow mug, 3" tall, marked "Shamrock"; an ivory-colored mug, 3" tall, marked "Eagle"; a dark green mug, 3" tall, marked "Deka"; an ivory-colored mug, identical to first.

Translucent Polystyrene mugs and tumblers, all 3.75" tall. (Top row, left to right) Light blue, marked "Aladdin Plastics L.A. Ca."; pink and yellow, marked "Allied Plastics"; yellow. (Bottom row, left to right) Red; red, marked "Plasmetl"; cobalt blue; green, marked "Allied Plastics."

Tumblers. (Top row, left to right) Pink, yellow, turquoise, and white, 4" tall, marked "Crystalon"; turquoise and white, marked "Crystalon," 3.5" tall; bright yellow and dark green, Polyethylene, marked "Shamrock," 3.5" tall.

Salmon red, lemon yellow, and turquoise tumblers in Polyethylene, 5" tall, marked "Country Craft by Proven Products." Also pictured, a pink plastic whisk and a glass pitcher with a fish motif.

Polyethylene mugs and tumblers. (Left to right) Dark green, 3.5" tall; pink, 3.5" tall; rosted blue, 3.5" tall; yellow, 3" tall; frosted red, 3" tall.

Polyethylene mugs and tumblers. (Left to right) Avocado and white, 6.5" tall, marked "Blisscraft of Hollywood"; ivory-colored, 6" tall; avocado, 5" tall; yellow, 4.75" tall; red, 3.75" tall; light green, 3" tall.

Polyethylene tumblers. (Left to right) Frosted yellow and red tumblers, 3.5" tall; copper-colored and mustard yellow tumblers by Stanley Home Products, 3.5" tall.

Four spangled Polyethylene tumblers in frosted, blue, blue-green, and yellow, 4" tall.

Spangled Polyethylene tumblers in frosted and blue, both 4" tall, yellow, 4.5" tall, and pink, 3.5" tall.

Three spangled Polyethylene tumblers in frosted turquoise, 3.5" tall, dark green, 3" tall, and frosted green, 3" tall.

Three pink Polyethylene mugs, 3", 2.75", and 3.5" tall.

Turquoise, yellow and spangled turquoise mugs in Polyethylene by Westland Plastics, 3" tall.

Green Polyethylene thermos mug, 3" tall; frosted blue spangled mug, 3.5" tall.

Mugs and tumblers. (Back row, left to right) Tumblers in red and black, marked "N"; large brown 'mix-up' tumbler; two tumblers with stacking grooves, marked "U.S. Halsey Inc." (Front row, left to right) A green Urea mug; a set of three Urea tumblers; an ivory-colored Melamine mug, marked "Melaware."

NAPKIN HOLDERS

Napkin holders were used to keep paper napkins in a neat stack on the table. The demise of paper table napkins because of their environmentally wasteful nature has led to the gradual obsolescence of napkin holders. Back in the 1950s, no one cared and these items were found in most kitchens. The plastic ones are, unfortunately, rather uninteresting, at least compared to the wooden figural and souvenir ones of earlier times. Still, some of the pierced holders in stylized floral and animal shapes are fun. Sadly, many of these fragile Polystyrene beauties cracked or broke from heavy use.

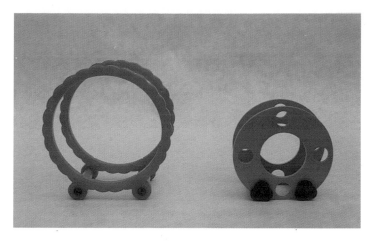

Napkin holders in butterscotch and black Catalin. *Courtesy of Cat Yronwode.*

Polystyrene napkin holders. (Left to right) Turquoise and clear, marked "Deka"; yellow with decals; mottled white and orange.

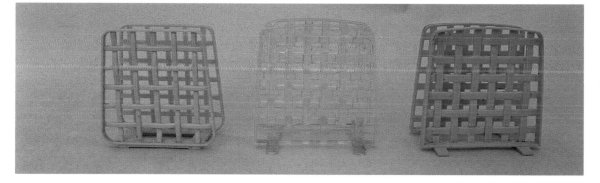

Three Polystyrene napkin holders in pink, clear, and green 'weave' designs.

Napkin holders. (Left to right) A yellow Polystyrene daisy design; a red Polystyrene 'weave' design; a Polyethylene design from the St. Labre Indian School.

Polystyrene napkin holders. (Left to right) Frosted pink; chartreuse with a pierced tulip design; orange with a pierced floral design, marked "Lustro-Ware."

NAPKIN RINGS

The collectible plastic napkin rings you will see here date from about 1935 to 1942 and are made of Catalin. Most were made by the Ace Novelty Co. In fact, these charming creatures were quite popular in their time. Many come shaped as figures of animals. These are already considered *very* desirable, and the prices reflect that! Don't expect to find these items at the bottom of bins in thrift stores. More likely you will be purchasing them at collectible shops. Again, the pieces shown are from the incredible collection of Cat Yronwode.

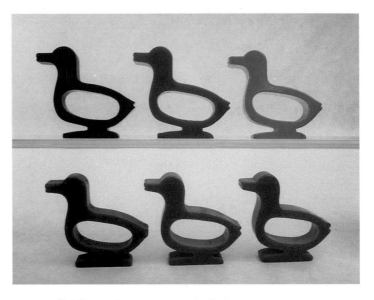

Catalin napkin rings in a duck design.

Catalin napkin rings, in a robin design.

68

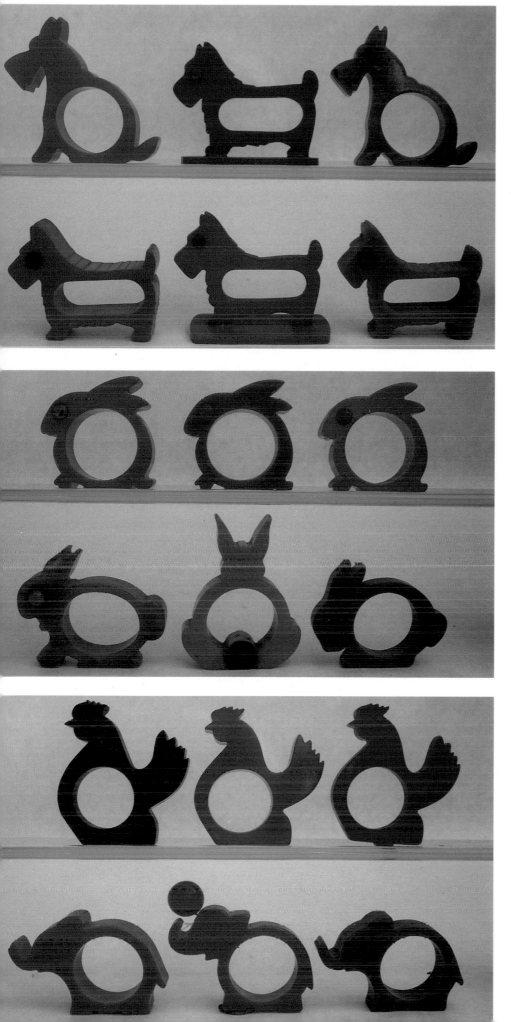

Catalin napkin rings in a scotty design. (Remember, scotty dogs were very popular in the 1930s because of Fala, President Roosevelt's Scotch terrier.)

Catalin napkin rings in a bunny design. The last two on the bottom row are European.

Catalin napkin rings in rooster and elephant designs.

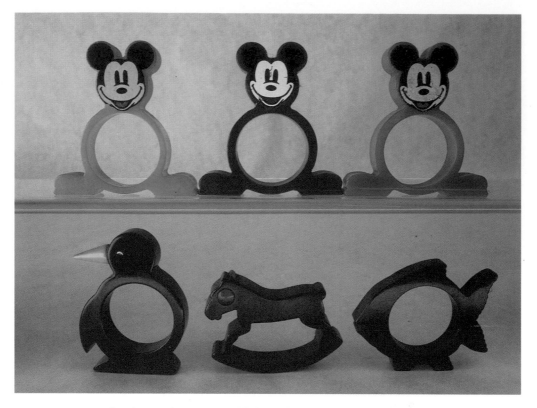

Catalin napkin rings in Mickey Mouse, penguin, rocking horse, and fish designs. The Mickey faces are pasted on, and there are reports of an early Bugs Bunny napkin ring!

An advertisement showing "gifts of cheery Plas-Tex," including informal dining sets and tumblers in "warm California pottery shades." From *Women's Home Companion*, 1952.

PICNIC ITEMS

Original plastic picnic sets of the 1930s and 1940s came in suitcases with compartments for food. The suitcase sets can still be found, but not always complete. Early table pieces are usually Urea with Catalin-handled flatware. A 1946 ad shows a picnic set with all the pieces made of green Urea. A typical set may include regular size plates, cups and saucers, mugs, food containers, and a thermos. By the 1950s most sets, including the utensils, were made of Polystyrene. Eventually, the suitcases were eliminated, and just the pieces were offered. These lightweight, brightly-colored and somewhat disposable pieces gave picnicking a whole new lift. 1950s picnic sets included the well-known divided plate, mugs or cups and saucers, bowls, flatware and sometimes other utensils, and a small salt and pepper set.

There were many divided plates many by various companies in bright, 1950s contemporary colors. It would be a misnomer to call these items picnic pieces. The sweeping trend towards all types of 'informal dining', even indoors, would have called for these items.

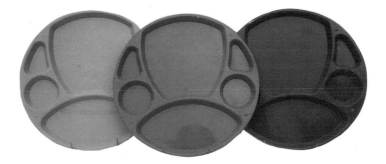

Picnic items, including 13" divided plates in gray, turquoise and salmon Polystyrene, marked "Plas-tex Corp."

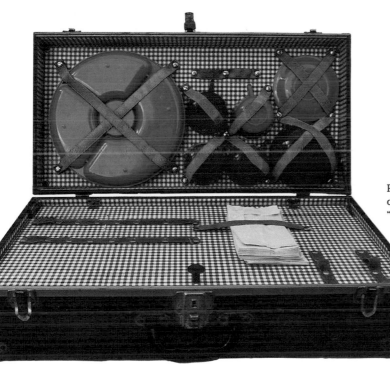

Picnic items and a picnic suitcase. The Polystyrene red and green divided plates, cups and saucers, and bowls are marked "Hemcoware."

Polystyrene picnic items. (Top row, left to right) Blue, yellow, and red divided plates, marked "USA." (Bottom row, left to right) Ivory-colored Plas-tex divided plate; turquoise divided plate and cup, marked "Lustro-Ware."

Polystyrene picnic items. (Top row, left to right) Red and pink divided plates marked "Regaline"; a red cup by Aladdin; a yellow cup by Arvin. (Bottom row, left to right) Yellow and green divided plates with matching flatware. Also pictured: turquoise, ivory-colored, and white thermal jug.

Square-shaped Polystyrene picnic cups marked "Maherware," in gray, jungle green, red, and chartreuse.

Polystyrene picnic items. (Top row, left to right) Green, blue, and red 8" plates marked "Dapol Plastics Inc." (Bottom row, left to right) Yellow and red cups and blue saucers marked "Plas-tex Corp."; bowls, 5" and 3.5" wide, marked "Burrite"; a blue mug marked "Plas-tex."

Polystyrene picnic items by Colonial Plastics Manufacturing Company. Round and square divided plates come in a good selection of colors, including pastels.

Polystyrene picnic items by Plastic Products Company of Hollywood, CA, including light blue, red, yellow, and green divided plates and cups, and an ivory saucer with a dark blue cup.

Refrigerator dishes. (Left to right) A yellow Blisscraft bowl with a frosted lid; a yellow Polystyrene dish with a clear top, 4" x 10"; a round, yellow, tumbler-shaped Polyethylene dish with a lid, marked "Republic."

Pink and yellow Polystyrene leftover dishes marked "Lustro-Ware."

SALAD UTENSILS

Salads, once a course between the entree and dessert, made a big comeback in the post-war world. Moved up to first course, salad fit in nicely with the new informal style of dining. Tossed salads were the salad of choice, considered the man's domain and de rigeur as an accompaniment to the ever-popular barbecue. The sets and utensils shown were made for this purpose. Any large bowl could serve as a salad bowl, but since salad bowls were originally wooden you can expect to find plastic examples in simulated wood textures. Salad utensils came in pairs and consisted of a large fork and spoon (usually about 12" long). A whimsical utensil frequently found (and still made) is called a 'salad scissors'. It is shaped like a large scissors with one tong the fork and the other the spoon. Some are shown in this section. The pale green salad scissors is made of an older plastic.

Salad utensils. (Left to right) A black and white fork; a pink and black fork; a turquoise and black spoon; a brown fork; a dark green fork, marked "Flintwood." All are made of Polystyrene except the Flintwood fork.

Polystyrene salad utensils: a pink fork marked "Hoffman Industries"; a turquoise fork; an ivory-colored spoon; a white utensil with a wooden handle; a white spoon with a red handle.

Acrylic salad utensils. (Left to right) A fork with a wooden handle; a fork with a metal handle; a spoon; an amber fork with a metal cinch.

Salad utensils. (Left to right) Salad scissors in opaque green (marked "Flexware"), translucent yellow, and translucent pink; a green nylon fork by Foley; a red Melamine fork by Mepal.

Salad utensils. (Left to right) A chartreuse set marked "Styson 1952"; a black set with silvered handles; a chartreuse set with black handles. The last two sets are the same design.

Salad utensils. (Left to right) Black agatized wood set; Raffiaware
Polystyrene set; black polystyrene set with silvered handles.

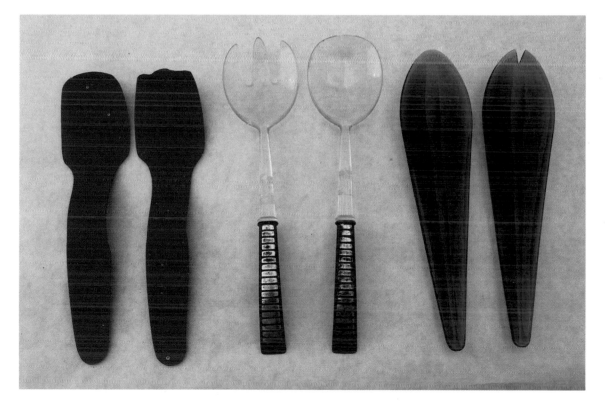

Salad utensils. (Left to right) Fluorescent pink acrylic set; clear
Polystyrene set with black handles; avocado acrylic set.

SALT AND PEPPER SETS

Shelves full of books and reams of paper have been devoted to salt and pepper sets. The sets shown here are only a sampling. Plastic lent itself readily to these sets and imaginations were stretched from the practical and inventive to the weird and whimsical. Figural and advertising sets are the most desirable, though the design element of many 'modern' sets is attractive.

Catalin salt and peppers. (Top row, left to right) Butterscotch/orange mushrooms; butterscotch/green mushrooms. (Bottom row, left to right) Brown/black mushrooms; orange and red balls on a chrome tray. *Courtesy of Cat Yronwode.*

Catalin salt and peppers. (Top row, left to right) A green set and a red set, both ball-shaped. (Bottom row, left to right) A coffee pot set with red handles, and a square butterscotch set. *Courtesy of Cat Yronwode.*

Catalin salt and peppers. (Top row) Butterscotch and green caddies with glass shakers. (Bottom row) A butterscotch, green, and red set with metal tops. *Courtesy of Cat Yronwode.*

Pairs of semicircular Catalin salt and peppers in green, red, and butterscotch. One of the green pairs has a tray. *Courtesy of Cat Yronwode.*

Salt and peppers. (Left to right) A turquoise and white egg-shaped set with a metal holder, and a metal burro with Polyethylene holders marked "Tupper!"

Polystryrene salt and peppers. (Left to right) A turquoise Lustro-Ware pepper; a small red Lustro-Ware set; a coral and white Lustro-Ware set; a red top and white bottom set; a red Lustro-Ware pepper. *The small red set and the red and white set are courtesy of Old Town Antique Market, Portland, Oregon.*

Salt and peppers. (Left to right) A red and white Polystyrene set, marked "Plas-tex Corp."; a yellow and red Sonette set made of Urea, with its box; a yellow dual shaker with metal spouts, marked "Hyland Prod. Company Portland, Or. pat 1940."

Salt and peppers. (Left to right) A light blue Urea shaker; a light green Celluloid shaker, marked "Carvanite"; a blue and white Urea set, marked "Carvanite"; a green and red bullet Urea shaker set, marked "Pachmayz, LA"; a sea-green Celluloid set.

The "Lovebirds" and "Rose Trellis" salt and pepper sets. *Both courtesy of Modern Times, Portland, Oregon.*

Salt and peppers. (Left to right) A blender set; coffee pots; a toaster set. *The coffee pots and toaster sets are courtesy of Modern Times, Portland, Oregon.*

Salt and peppers. (Left to right) Millie and Willie penguin set (advertising characters for Kool cigarettes; Uncle Mose and Aunt Jemima set (advertising characters for pancake mix). Both sets marked "F&F Mold and Die Works, Dayton, Ohio." There are many other plastic Aunt Jemima items. *Millie and Willie courtesy of Modern Times, Portland, Oregon.*

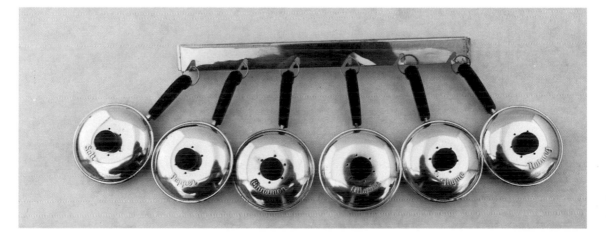

A six-piece salt and pepper and spice set shaped like frying pans hanging on a metal rack, marked "Brillium." *Courtesy of Modern Times, Portland, Oregon.*

Polystyrene salt and peppers. (Left to right) Autumn gold and white with a decal; white and red with butterflies; ivory-colored and white set marked "Max Klein Inc."

Polyethylene salt and peppers. (Left to right) A corn set, a yellow and clear set, a turquoise set, and a St. Labre Indian School set.

Three Polyethylene salt and pepper sets in the same shape but with different decals and colors.

Salt and peppers. (Left to right) Blue and white Polyethylene marked "Steri-Lite"; a black pepper grinder; yellow Polystyrene.

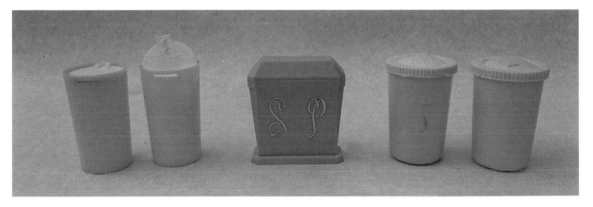

Salt and peppers. (Left to right) A yellow and white Polyethylene set, marked "Remembrance, Brown and Bigelow, pat. 1962"; a turquoise Polystyrene dual shaker; a yellow and white Polyethylene set.

Salt and peppers. (Left to right) A metallic Polystyrene set marked "Admiration"; an avocado and white Polyethylene dual shaker; a turquoise and clear Polystyrene set (in wrapper), marked "Steri-lite, United Plastics Company Fitchburg, Mass."

Besides the standard scoops used in stores and at home for scooping dry goods (flour, sugar, etc.) I have included ice cream scoops and coffee scoops, the latter still being produced as premiums. Coffee scoops (or coffee measures) usually have advertising printed or embossed on them or simply say something like "one full scoop equals one measure of coffee." Any type of scoop with increments on it will be listed under "Measures."

(Left to right) Metal scoops with plastic handles in white and yellow, marked "Bonny Corp."; turquoise, pink, and yellow Polystyrene ice cream scoops.

(Left to right) Turquoise Dip 'n' Flip scoop; a small blue scoop; yellow and brown scoops marked "Made in Hong Kong"; blue and pink Polystyrene scoops marked "Cal. Plastic S.F. Ca."

Scoops. (Back row, left to right) A Polyethylene ladle; turquoise and avocado Polystyrene scoops (in bowl), marked "Lustro-Ware." (Front row, left to right) A yellow Polyethylene scoop; a red Polystyrene scoop marked "Safetyware"; a turquoise advertising scoop for Admiral TV, Portland; a white advertising scoop from Valley Isle Bamboo, Wailuku, Hawaii; a red Polystyrene scoop. Also pictured are a pink metal flour canister with a metal top, a blue Blisscraft bowl, and a hot pad in pink and black.

Coffee scoops. (Left to right) A red scoop marked "B&B"; a yellow scoop with a Folger's advertisement; a metal scoop with a red handle; a yellow Fuller premium scoop; a small yellow scoop.

Polystyrene spoons. (Left to right) A four-piece white and blue spoon set; a three-piece spoon set in a ceramic wall holder; a three-piece spoon set in yellow, red, and white.

SERVING BASKETS

Also called 'bread baskets'. I remember my mother serving sliced bread or rolls in these at the dinner table. If warm, the bread or rolls were folded in a napkin. These baskets were also popular for use in restaurants and malt shops where customers received burgers or fried chicken ('chicken in a basket') in them. The polyethylene baskets were more suited to the constant institutional use of restaurants. The standard criss-cross pierced design is called a 'basketweave'. The weaving on Polystyrene baskets is prone to breakage.

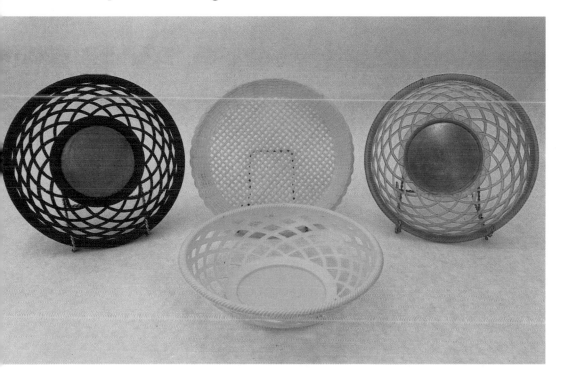

Four serving baskets. The three large examples (translucent red with a metal bottom, translucent green with a metal bottom, and opaque yellow) are Polystyrene, marked "Plasmetl." The low, opaque, smaller basket in the center is made of yellow Polyethylene.

Various round serving baskets in Polyethylene.

Unusual woven plastic pieces made in Hong Kong, including a yellow oval basket, an orange round basket, and a small yellow nut cup, marked "Made in Hong Kong."

Serving baskets in pink and salmon red Polyethylene.

Polystyrene serving baskets. (Top row) Red and yellow with daisies in the basket. (Bottom row) A pink basket with a hand-painted rose, marked "Plasmetl"; a white basket; a metallic bronze basket marked "Plasmetl."

SERVING PIECES

The 1954 edition of *The Snack and Party Cook Book* shows a beautiful Melmac platter and advises: "Plastic platters and trays are highly durable and long lasting and come in a wide variety of colors. Not for cooking, just for serving your hors d'oeuvres." Serving pieces, however, are not limited to hors d'oeuvres platters. This is a catch-all section encompassing various serving pieces, from bowls and trays to snack sets and cream/sugar sets.

Acrylic was used extensively for serving pieces. Since acrylic is the most scratchable of the plastics listed here, finding pieces in excellent shape is rare. You will also find many serving pieces with textural surfaces mimicking other materials. There is little known about the translucent fiberglass pieces embedded with leaves, butterflies and gold spangles. I have seen other items made of this material besides those shown here, including a huge round serving bowl. Some pieces have actual leaves and butterflies in them, while other pieces have decals instead.

SILEX Selects WATERBURY PLASTICS

Silex cocktail tray molded of scarlet Durez by Waterbury Plastics

A Silex cocktail tray molded by Waterbury Plastics of Waterbury, CT.

(Left to right) A metal server with a red and butterscotch center handle, 10", a chrome serving tray with red Catalin handles, 11.5"; a metal server with a green and butterscotch center handle, 8". *Courtesy of Cat Yronwode.*

(Back row, left to right) A metal covered dish with red Catalin handles and finial, marked "Keystonwear," 7.5"; a metal casserole holder with butterscotch Catalin handles. (Front row) A metal serving tray with butterscotch handles and a matching fork with a ball knob, marked "Manning Bowman," 4" x 9". *Courtesy of Cat Yronwode.*

Toast n' Jam set molded of Melmac by Whitso, Inc. in Schiller Park, IL. Jars are sand-colored and gray with tops of Persimmon Orange and Chocolate Brown. The tops also come in Midday Blue and Terra Cotta Red. The toaster is a McGraw Toastmaster. From *Modern Plastics Magazine*, 1954.

(Left to right) A two-tier brass serving tray with butterscotch Catalin handles; matching crumb sweepers. All made by Chase. *Courtesy of Cat Yronwode.*

Celluloid and Bakelite holders for the small pencils used for bridge games. These really choice 1930s pieces are shaped like a scotty, a magpie, and an elephant. *Courtesy of Cat Yronwode.*

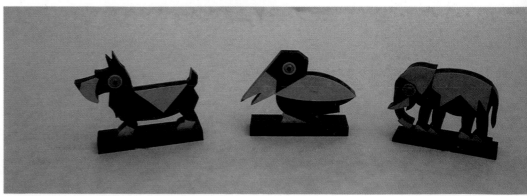

Red and white Polystyrene serving sets, each including creamer, sugar, shaker(s), and tray. Both made by Federal Tool Corp. of Chicago, IL.

Two large, unmarked, swoop-shaped acrylic bowls, streaked orange and clear, and white and gray.

(Back row) An ivory-colored styrene bowl with metal "feet," 10". (Front row) Black bowls, 6", marked "Styson Art Products Co."; a low copper-colored bowl, 14", marked "Bolta - General Tire."

(Back row, left to right) A Polystyrene salad set that includes a large turquoise salad bowl, four yellow dishes, and utensils, marked "Hofmann Industries Inc,. Sinking Springs, Pa."; a sea-green salad set made of styrene. (Front row) Nut dishes in styrene, marked "Delagar"; combination coaster/dish sets in blue and red styrene, marked "Hofmann Industries Inc."

Fiberglass-type serving pieces with embedded objects. (Left to right) A 12" x 17" tray with decal butterflies; a 9" x 14" tray with real leaves and mica; a 6" x 16" tray and a 6" bowl with gold confetti and spangles.

SHAKERS

Also known as 'dispensers'. The majority of shakers were for sugar. Some sugar shakers were part of a set including salt and pepper shakers. Polyethylene squeeze bottles for condiments are pictured, including figural ones. Spice jars and cleanser dispensers are also included here. This section also includes glass dispensers with plastic tops, which, of course, replaced metal tops. You will also find metal and plastic tops combined.

(Left to right) A marblized pink Polystyrene dispenser; a salmon and white Cellulose acetate dispenser; a red Urea dispenser for Babo Cleanser (the can fit inside) with a rubber bottom.

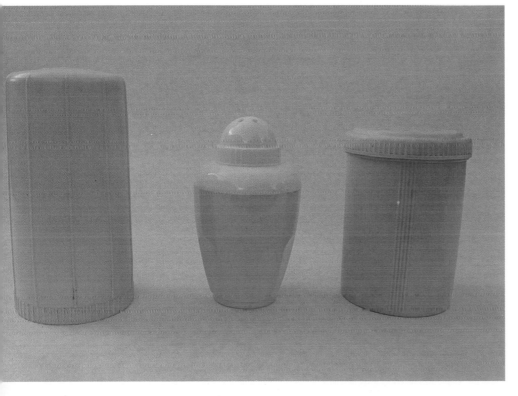

(Left to right) A yellow Polyethylene dispenser marked "Made for Babo Cleanser"; an aqua and white Polystyrene shaker; an aqua and white Polyethylene shaker.

(Left to right) A Lustro-Ware sprinkler; a frosted and turquoise Polyethylene sprinkler; "Mr. Sprinkle" sprinkler in purple and white Polyethylene; a clear sugar dispenser (bottom-marked "Eagle") with a copper-colored top; a yellow and white Polyethylene sugar dispenser marked "Remembrance."

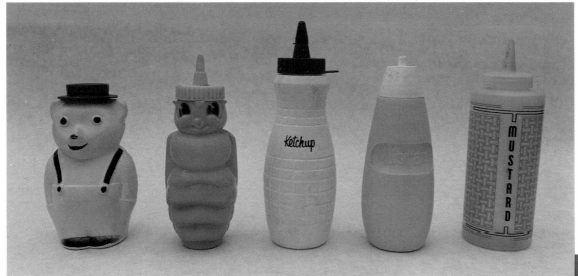

Polyethylene shakers and dispensers. (Left to right) A 1950s-era Squeezit bear with a Polystyrene hat; a Squeezy Bee; a ketchup dispenser in white and red; a yellow and white mustard dispenser; a frosted mustard dispenser with printing.

Shakers and dispensers with red and green Catalin Handles.

Shakers and dispensers with plastic tops. (Left to right) A yellow and white nut chopper marked "Gemco"; a turquoise and white nut chopper marked "Federal Housewares"; a yellow sugar marked "Gemco"; a red dispenser marked "G. C. Co."

Shakers and dispensers with plastic tops. (Left to right) A light green salt and pepper set; a green Cellulose Acetate sugar dispenser marked "Measuring Device Co."; a pink pitcher marked "EZ Pour Corp."; a yellow and red salt and pepper set (possibly new).

A four-piece set of dispensers (salt and pepper, sugar, and syrup) with matching blue plastic elements, marked "Hazel Atlas Co."

Shakers and dispensers with plastic tops. (Left to right) One in red Catalin and metal; three in red Polystyrene. *Polystyrene pieces courtesy Old Town Antique Market, Portland, Oregon.*

Shakers and dispensers with plastic tops. (Left to right) A three-piece accent set with red tops and a plastic holder; a lone red salt. *Accent set courtesy of Old Town Antique Market, Portland, Oregon*

Silverware Trays

Also called a 'cutlery tray' or 'divider', the standard tray has three or four horizontal compartments and a short vertical one. The horizontal compartments were for forks, knives and spoons and the small section was for miscellaneous items. This category is one of the classic examples of how plastics replaced another time-honored material. These trays had been forever made of wood. Credit Columbus Plastics Products, Inc. for introducing the Polystyrene silverware tray in the late 1940s. This Lustro-Ware item was an instant success and quickly relegated the wooden versions to obscurity. Many other plastic companies made these trays. Most of these trays ended up in workshops and garages. Expect to find the trays especially the Polystyrene ones — extremely dirty or cracked. I have seen the 1950s tray with gold spangles (pictured here) in other colors as well.

Polystyrene silverware trays. (Top to bottom) Red, marked "Nudell Plastics"; green and white marblized; red, marked "Lustro-Ware."

Silverware trays. (Left to right) White with flatware designs in gold; yellow with gold spangles; mustard yellow Polyethylene marked "Blisscraft of Hollywood."

Silverware trays. (Left to right) Aqua Polystyrene, marked "Glama-Ware, Vermont Plastics"; pink Polyethylene; green Polyethylene marked "Plas-tex."

UTENSILS

Another vast category. What is the difference between a utensil and a gadget? A utensil is defined in Websters Dictionary as "an implement . . . used domestically, as in a kitchen." A gadget is defined in the same dictionary as "a small specialized mechanical device." Essentially, utensils are non-mechanical tools. Whisks, spatulas, tongs, stirrers, etc. all fit into this category. Many early examples have metal bodies with plastic (usually Catalin) handles. The large selection featured here is from the collection of Cat Yronwode. Examples from later eras are all-plastic, and like flatware run the gamut of plastics from that time. Nylon utensils were popular, and were used in conjunction with Teflon cookware — both still being made today.

Catalin-handled cake breakers. The word from the Catalin world is that cake breakers are the most common utensil available today. Many have reported finding them in their original box. Well, how often do you think they got used?

Catalin-handled cheese slicers.

Catalin-handled strainers.

Catalin-handled utility knives. The second from the left is marked "Federal"; the fourth "Marvel"; the fifth "Sheffield"; and the sixth "Lifetime Cutlery."

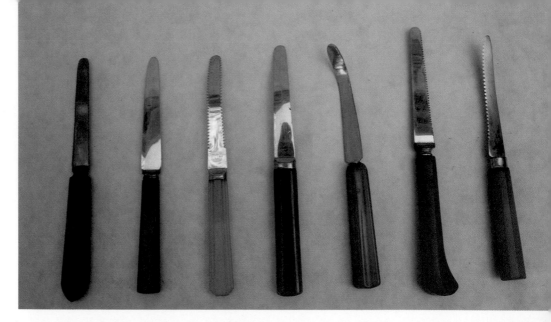

Catalin-handled grapefruit knives.

Large utility knives with Catalin handles, showing the range of colors called "butterscotch" to be found in Catalin. The third knife from the left is marked "Sheffield England"; the sixth is marked "Valley Forge."

Large utility knives with Catalin handles in black, red, and ivory. The first on the left is marked "StainPruf"; the third from the left "Sheffield"; the fourth "Quikut"; the sixth "Newbridge."

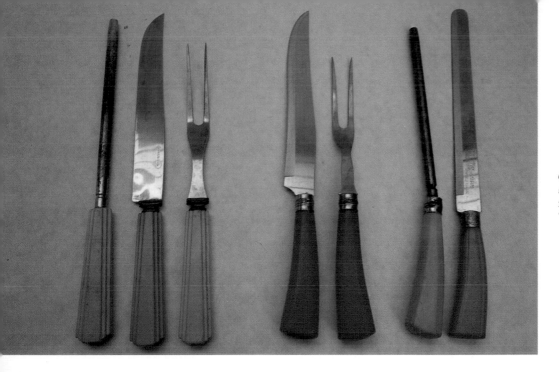

Catalin-handled carving pieces. The second from the left is marked "Washington Forge"; the third is marked "Sheffield."

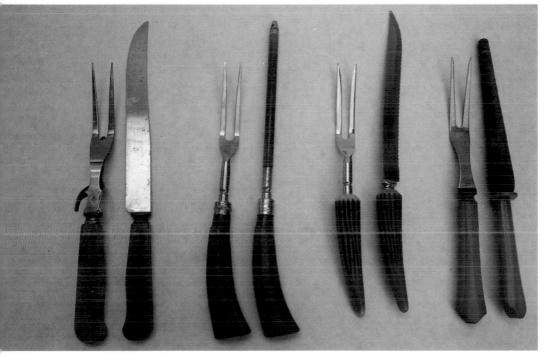

Catalin-handled carving pieces. The first on the left is marked "StaBrite."

Catalin-handled carving pieces.

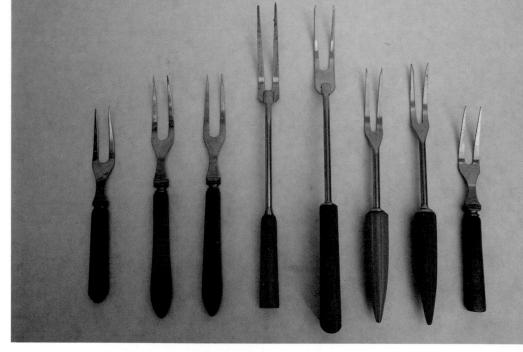

Catalin-handled utility forks. The fourth and fifth from the left are marked "A & J"; the sixth and seventh are marked "Androck."

Catalin-handled steak knives. The second and fourth from the left are marked "Henry's."

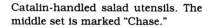

Catalin-handled salad utensils. The middle set is marked "Chase."

Catalin-handled iced tea spoons.

Catalin-handled utility spoons. (Left to right) Marked "Corona," "Androck," "Corona," and "Androck."

Catalin-handled long spatulas. (Left to right) Marked "Androck," "Enlgishtown," "Androck," and "Androck."

Catalin-handled whisks. The first from the left is marked "Easy Aid"; the second "A & J"; the fourth "Ecko Eterna."

Catalin-handled pierced spatulas. (Left to right) Marked "Androck," "Englishtown," "Ecko Eterna," and "Vollrath."

Catalin-handled mashers. The third from the left is marked "Androck," the fourth "Ecko Eterna."

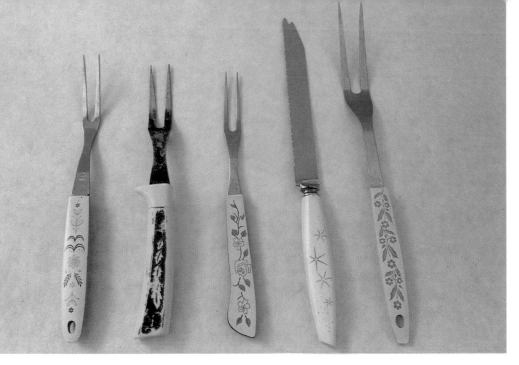

Melamine-handled meat forks, and a carving knife with stylized decorations. The knife is marked "Sheffield."

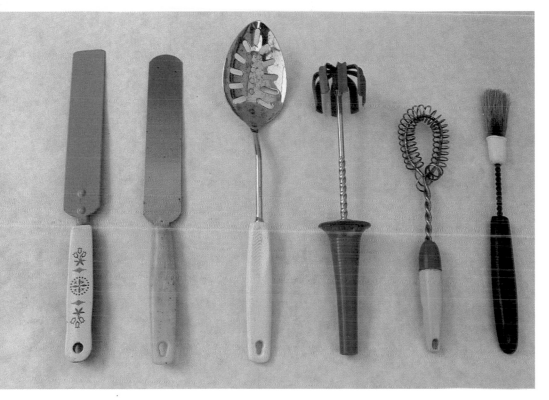

(Left to right) A long, thin, Melamine-handled spatula; a yellow Polyethylene-handled frosting knife; a white-handled slotted spoon; a whip with an avocado-colored handle; a yellow- and green-handled whisk; a black-handled pastry brush.

(Left to right) Three Polystyrene pie or cake slicers in turquoise, ivory and red; a green acrylic knife; white and mustard-yellow Polyethylene pie or cake servers. I have seen the shape of the first two slicers in metal – that mold really gets around!

Polyethylene utensils. (Top to bottom) A red slotted spoon marked "Foley"; a yellow whisk marked "Stanley"; a red whisk; a red scraper.

Polystyrene planters molded by Quality Molding Company in 15 styles and 20 different sizes, not to mention a variety of colors. From *Modern Plastics Magazine.*

VASES AND PLANTERS

Another category which can also be a household item. Cheery kitchens called for flowers and plants. Red geraniums and ivy were two plants whose popularity started in the 1940s and persisted for decades. You will find these two motifs on everything from curtains to wallpaper. The smaller vases and planters pictured here seemed to fit into the kitchen schemes of the times. The Polystyrene planters of this period quickly replaced the standard clay pot. Expect to find many more examples. The vases marked "Fronzwood" appear to be some type of fiberglass material.

A tall, footed, dark green vase, with a paper label that reads "Fronzwood"; an ivory-colored planter; a yellow vase; and a low, rectangular, dark green planter. These pieces all have a strange fiberglass-like texture. It would be wonderful if more pieces in this line were found.

(Left to right) A white pierced planter; a Lustro-Ware flower holder; a square, yellow Lustro-Ware flower pot.

(Left to right) A pink planter with a black base; an orange planter with a black base; a round pink planter. The pink planters are marked "Quality Molding Co."

WALL-HANGINGS

These items include decorative objects and planters. Remember all those plaster wall-hangings from the 1930s and 1940s of Dutch children, fruit, cartoon vegetables and fish? I haven't found any plastic versions, which I assume would have been made to supersede the easily damaged plaster ones. The pieces found so far seem to be styled after the popular 'wrought iron as art' look of the 1950s.

Two wall hangings, a Dutch boy with planter baskets, and a chef holding a cake.

Two wall-hangings: one a metallic bronze color with an ivy and circle design, and the other in pink with black Polystyrene. The hooks hold little planters on chains.

Rooster wall hangings.

A two-piece black horse and buggy wall hanging, with translucent white planter boxes.

WASTEBASKETS

Another 'but that's not kitchenware!' item. Remember, plastic replaced just about everything in the kitchen. These are not garbage pails but small utility wastebaskets. The development of reinforced and tougher Polyethylene in the 1960s facilitated the creation of heavy duty garbage pails and trash cans. The smaller utility baskets shown here were fashionably meant to sit near the door, phone station, or dining table. They are still being made today.

(Left to right) A pierced wastebasket in ivory-colored pierced Polyethylene, and another in dark green Polyethylene marked "Max Klein Inc."

116

(Left to right) A pink Polyethylene wastebasket with gold starbursts, marked "Lustro-Ware"; a yellow oval Polyethylene wastebasket with outer space motifs.

WATERING CANS

The items presented in this section are not the large garden-type watering cans, but the smaller ones used for watering houseplants. Houseplants were at one time strictly the domain of the kitchen (and sometimes the bathroom as well). Look for Polystyrene cans with stylized shapes and designs as well as backstamps. Beware of cracked bodies and broken spouts. Polyethylene watering cans came later.

Polystyrene watering cans. (Left to right) Red, in a conical shape; yellow and frosted, in a bulbous shape.

Polystyrene watering cans. (Left to right) Red, in a bulbous shape; yellow; green, in a bulbous shape; yellow, in a bulbous shape.

Watering cans. (Left to right) A metallic, copper-colored Polystyrene can in a streamlined shape; a turquoise Polyethylene can marked "Lustro-Ware."

COMPANIES AND MISCELLANEOUS KITCHEN ITEMS

COLUMBUS PLASTIC PRODUCTS — COLUMBUS, OHIO

I hold a particular passion for this paramount of Poly-practicality. The items produced for the kitchen by this company could probably fill a book on their own. The company's contribution to the plastic kitchenware industry is unparalleled. Started in 1938 as a producer of plastic components for industry, Columbus began trying their hand at molded plastic housewares. By the end of the war, their products were gathered together under the famous Lustro-Ware line. By the late 1940s, Columbus was recognized as having the most complete line of thermoplastic housewares in the country. NOTE: *Thermo-*

plastics are plastics that can be re-heated and re-formed, like Polyethylene and Polystyrene. *Thermo-setting plastics*, including Melamine, are permanent and heat resistant once they are heated and formed.

Columbus Plastic Products was the first company in its field to have an in-house production design staff. They also pioneered many innovations in the engineering and production of plastic items. However, most desirable for the collector are their breakthrough items for the kitchen. By the mid-1960s, Columbus Plastic Products had over three hundred items in its Lustro-Ware line! At this time, the company was acquired by the Borden Company. Borden continued to produce the Lustro-Ware line until, sadly, it was discontinued around 1982.

Today, Lustro-Ware items are plentiful, which makes them great fun to search for and collect. Many items found are Polystyrene, including the well-known and easily recognized canisters (see the canister section). You will discover household items for cleaning as well as kitchen-related wares. Most items you will find will be from the 1950s. As mentioned, the Lustro-Ware line ran until 1982 but I have yet to find any items marked "Lustro-Ware" from this later period. Lustro-Ware items are identified on the back and all items have a stock number. I always look forward to what new Lustro-Ware piece I will discover. Viva La Lustro-Ware!

Lustro-Ware canisters, made of Polystyrene. Red and yellow jars with white tops, and a tan sugar-size canister with a white top and a 'rooster' design.

Lustro-Ware canisters, made of Polystyrene. (Left to right) A white cookie jar; a coral flour canister with a white lid; a red canister with a white lid and a flower design.

Lustro-Ware canisters, made of Polystyrene. This four-piece set, off-white with white lids, features a 'rooster' design.

Lustro-Ware canisters, made of Polystyrene. A four-piece set, red with white lids.

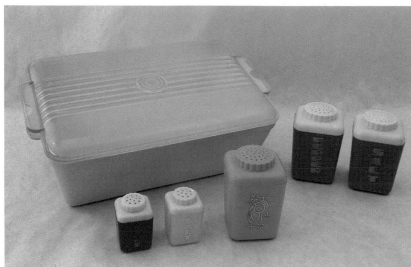

Lustro-Ware items, made of Polystyrene. A yellow and clear storage bin, small yellow and red salts, a turquoise pepper, and a red salt and pepper set.

Lustro-Ware's Polystyrene bread box, yellow with a white 'flip-up' lid. There is lettering on this piece, but it is faded to the point of invisibility.

Lustro-Ware's Polystyrene bread box, yellow with a white 'roll top' lid.

From the Lustro-Ware line, made of Polystyrene. (Back row) White
and pink pitchers. (Front row) Divided plates in turquoise, frosted
yellow, and chartreuse, and a turquoise cup.

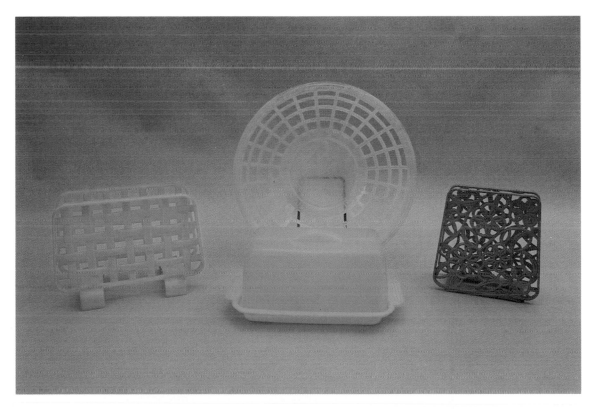

From the Lustro-Ware line. (Back row, left to right) A yellow napkin holder, a frosted serving basket, and an orange napkin holder,
all Polystyrene. (Front row) A yellow and frosted Polyethylene butter dish; this piece was also called a 'cheese dish'.

From the Lustro-Ware line. (Back row) Red and turquoise soap dishes with black inserts; a pink wastebasket with a swoop-shaped top. (Middle row) Pink and yellow soap dishes. (Front row) A turquoise and green flyswatter. All items are Polyethylene.

From the Lustro-Ware line. (Left to right) A yellow and frosted Polystyrene watering can, a yellow and clear Polystyrene sprinkler, and a turquoise Polyethylene watering can.

From the Lustro-Ware line, made of Polystyrene. Three silverware trays in yellow, red and marblized green.

BLISSCRAFT OF HOLLYWOOD

Blisscraft's slogan is "Always Look for the Crown of Quality." Their logo actually has a little crown in it! This ecstatic-sounding kitchenware line was manufactured in Gardena, California in the 1950s when Polyethylene was eclipsing Polystyrene as *the* kitchenware plastic. They were doing so well that by the mid-1950s they opened a new and larger plant. Most products found are Polyethylene, though a few Polystyrene pieces have been discovered. Blisscraft items all have molded scalloping somewhere on the edges of each piece. The original Poly-Pitcher pitcher set came in three sizes: 2 quarts, 1 quart and 1/2 quart. Mixing bowls come in nested sets of six and have been found in the following colors: bright orange, salmon red, deep blue, lime green, frosted, yellow, cobalt, brick red, sage green, peach, ivory, aqua and pink! Assembling a usable set is fairly easy.

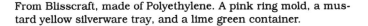

From Blisscraft, made of Polyethylene. A pink ring mold, a mustard yellow silverware tray, and a lime green container.

Blisscraft pitchers in four sizes. The pink lid has embossed gold fruit on it.

From Blisscraft. (Top row, left to right) A yellow and clear Polystyrene creamer and sugar set; an avocado and white Polyethylene mug; a red and clear Polystyrene creamer. (Bottom row, left to right) A copper-colored Polystyrene butter dish (lid missing); a yellow and frosted Polyethylene covered dish; a yellow and frosted Polyethylene butter dish.

Blisscraft's six-piece mixing bowl set, pictured here in orange, red, cobalt blue, lime green, frosted, and yellow.

CUT GLASTIC

Authentic antique cut glass was originally a product of Europe until the Industrial Revolution in America created factories, mostly on the East Coast, to produce domestic versions of this beautiful, high-quality glassware. By the turn of the century so much cut glass had flooded the market, the public's interest in it faded. At the same time pressed glass appeared as a less expensive substitute. Pressed glass has a more rounded feel if you run your fingers over the 'etching' in the glass.

Any avid glass collector would consider these plastic imitations pictured here 'cheap and tacky'. In truth, they are — but they have also served their purpose well, giving average people a chance to feel they were using elegant items in their homes, even if they knew it was only plastic. These imitation 'cut glass' items are extremely plentiful and still being made today, though they are no longer as popular. Some of the contemporary ones come in transparent vivid colors like yellow, orange, cranberry, and green! The older pieces are clear acrylic and can be identified by their unfortunate yellowing. Most pieces are unmarked but some pieces are backstamped "Regaline," a name that has been found on various other kitchen and household items.

Molded plastic compact made to imitate expensive cut glass. From *Modern Plastics Magazine*, 1946.

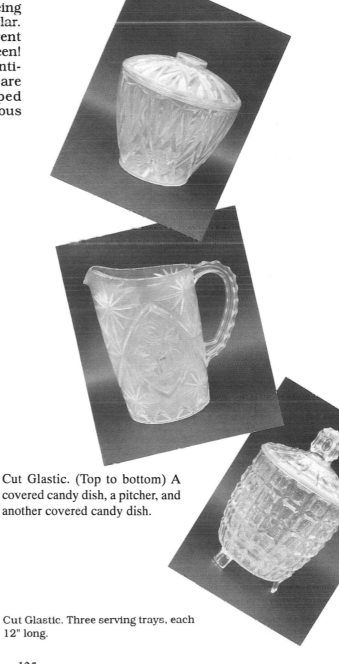

Cut Glastic. (Top to bottom) A covered candy dish, a pitcher, and another covered candy dish.

Cut Glastic. Three serving trays, each 12" long.

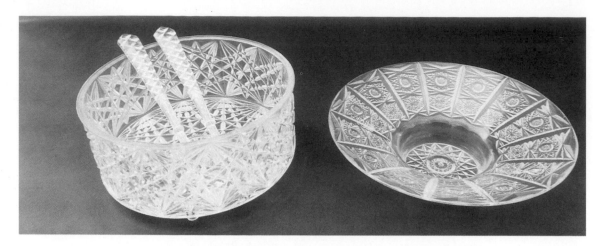

Cut Glastic. (Left to right) A footed serving bowl with salad utensils, and a wide-rimmed serving dish.

Cut Glastic. A tall creamer and sugar set (at left and right), a short creamer, and a three-sectioned handled serving tray.

Cut Glastic. (Back row, left to right) A napkin holder, a tumbler, a 9" vase, a tumbler, and a napkin holder. (Front row, left to right) A butter dish, a low serving bowl, and a butter dish.

Thermal Ware

For our purposes, a thermal object is defined as any kitchen or dinnerware piece that is double-walled, insulated and sometimes vacuum-sealed. This includes pitchers, tumblers, cups, bowls, mugs and various containers and serving pieces. You will find numerous pieces by many companies backstamped with a variation of Thermo (e.g. Thermo-Ware, Thermo-Frost, etc.) Thermal items became very popular in the mid-1950s.

The Thermo-Serv Company, a division of the West Bend Company, was the most successful thermal ware producer. Starting in the early 1950s, they created the easily recognizable black and gold thermal ware line used in restaurants. They were also the first to create silk-screened forms and to place printed paper inserts between the clear outer and solid inner walls.

Insulated tableware made of Styron (Polystyrene). From *Modern Plastics Magazine*, 1956.

Pictured in this section is a popular and easily recognizable thermal line. The line is unmarked, so I will refer to it as "Straw-weave." You can find many pieces of Straw-weave, but these Polystyrene items cracked and stained easily and the straw inside tends to 'unravel'. Other pieces of Straw-weave have been seen besides those pictured here, including an ice bucket and a thermos.

Thermal ware in the Straw-weave design. (Back row, left to right) Tumbler, 6" tall, chocolate; two tumblers, 5.5" tall, yellow and blue; tumbler, 6" tall, turquoise. (Front row, left to right) Bowls, 5" diam., with salmon and turquoise interiors, white Polyethylene lids lightly embossed with ivy vines.

Thermal ware in the Straw-weave design. Mugs, all 3.5" tall, in ivory, chocolate milk brown, salmon, light blue, yellow, and pink.

Thermal ware in the Sunfrost design. A mustard yellow pitcher; red and purple mugs, 5.5" tall; a blue tumbler, 5" tall; orange, yellow, avocado, and turquoise mugs, 3" tall; and a mustard yellow covered bowl with metal handles – possibly an ice bucket?

Thermal ware in the Bolero design, including orange, red, and green tumblers, 5" tall, and orange and blue bowls, 5" diam.

Thermal ware in the Olympian design. (Left to right) A yellow mug, 5.5" tall; a yellow custard cup, 2.5" tall; a red mug, 3" tall; a red tumbler, 6" tall. All pieces have a Greek-style border design lightly embossed below the white rim.

Thermal ware in the Cornish design. (Top row) Blue, turquoise, and yellow tumblers, 5" tall; peach and orange tumblers, 3.5" tall. (Bottom row) Salmon, yellow, turquoise, and orange mugs, 3" tall.

Thermal ware, miscellaneous. (Left to right) A salmon mug with chicken wire insert, 3.75" tall; a yellow tumbler, 5" tall; A white pitcher with a plaid paper insert, marked "Thermo-Serv by West Bend"; a mug with a daisy design paper insert, marked "West Bend," 6.5" tall; a salmon mug, 3.75" tall.

Thermal ware. Bowls in various colors, 5" diam., with white plastic ivy vine inserts; pink tumblers, 5" tall, with white starbursts.

Thermal ware. (Left to right) A handled jug, 8" tall, ivory and aqua; A thermos-type jug, 10" tall, ivory, turquoise, and white, marked "Bee Plastics Inc." and "It's A Honey!"

Thermal ware tumblers. (Left to right) Red and yellow, 5.5" tall, marked "Eagle"; pink with gray spangles, 6" tall; turquoise, 5.5" tall; avocado, 5" tall, marked "Gitsware."

A set of thermal ware bowls, including a 10" serving bowl in lime green, and four 6" bowls in lime green and pink, marked "Therm-O-Bowl designed by Reinecke." Also pictured are a metal TV tray, white with black and gold modern designs, and waxed paper beer cups.

Thermal ware. Mugs, 3" tall, in turquoise, pink, gold/black, and golden yellow, and a casserole, 9" x 9", in gold and black with a clear lid.

RAFFIAWARE BY THERMO-TEMP

This highly popular thermal ware line, made by the Mallory Randall Corporation, imitates the texture of raffia weaving using a thick Polystyrene. Raffiaware looks like it was made to be 'patio ware' for informal outdoor dining. Alas, many pieces appear to have wandered too close to the barbecue pit. Tumblers, cups and bowls are most easily found. Keep an eye out for those snack trays. They aren't marked "Raffiaware" (they are marked "Thermo-Temp") but are by the same company and obviously meant to hold the Raffiaware mugs. The casserole with the metal insert is a real find! So far, no plates, platters, or large trays have been found.

Raffiaware, including a turquoise pitcher, 6" tall mugs in pastel orange and deep blue, and 5" bowls in deep blue and salmon.

Raffiaware footed bowls, 4" diam., in light green, pink, light blue, green, salmon, chocolate milk brown, and deep blue.

Raffiaware, including a salad bowl, 10" diam., in avocado with a matching lid; a salad utensil set; a 10" covered serving bowl with an acrylic lid and a metal insert; and a deep blue bowl, 5" diam.

Raffiaware, including tumblers in pastel orange, deep blue, chocolate, and yellow, 5" tall; mugs in pink and blue-green, 3.5" tall, with matching snack trays.

Lustro-Ware

BLISSCRAFT of HOLLYWOOD

U.S.A.

by ThermoTemp™

Eyes are on tableware molded of MELMAC*

Eyes are on tableware molded of Melmac. From *Modern Plastics Magazine*, 1948.

A post-war Melamine set offered to the public by Northern Industrial Chemical Company. From *Modern Plastics Magazine*.

Chapter Four
Plastic Dinnerware: Gone But Not Forgotten

Prior to World War II, only one plastic showed any promise as dinnerware — Urea-Formaldehyde, developed by the British in 1924. The American Cyanamid Company bought the rights to produce Urea and greatly improved it in the 1930s. At this point there were few major dinnerware manufacturers willing to produce their products in Urea plastic. Even with improvements and fillers, Urea had *major* drawbacks as dinnerware material: though heat- and alkaline-resistant, it was not acid-resistant. It swelled and warped when immersed in water, and cracked easily. It was apparent Urea was not the right material to be used for large scale manufacture of dinnerware.

But war clouds were approaching. Even before America's entry into World War II, the country was gearing towards war production. The armed forces needed a new plastic to line steel helmets and dinnerware which could withstand the rough treatment of war usage. At this time researchers at American Cyanamid 'rediscovered' Melamine - formaldehyde. In 1937, tests revealed that Melamine was unique in its resistance to breakage and wear, had a hard, scratch-resistant glossy surface, absorbed little water, and was odorless and tasteless (chemically speaking). It was obviously superior to Urea for dinnerware and was named Melmac.

When Melamine was released for civilian use after the war, plastic dinnerwares were still used chiefly in institutions and their full potential had not been explored. In 1946 there were only two companies producing plastic dinnerware lines (compared with almost twenty companies ten years later). However, manufacturers were quick to pick up on the trends of the times. Plastic dinnerware was regarded as new and daring — a break from tradition. Modern dining called for sturdy, inexpensive and attractive dinnerwares. Manufacturers stressed Melamine's qualities: it didn't break like china or glass, it didn't scratch (though this quality in particular turned out to be untrue), it was lightweight, versatile and of high quality. This appealed to Baby Boom-era parents saddled with an oversupply of rambunctious tots and restless teenagers. Melamine definitely fit the bill. These selling points were also important because Melamine dinnerware was not cheap when first introduced. Its prices were comparable or even slightly more expensive than traditional china. Salesmen assured housewives that the cost would be offset by Melamine's durability and longevity.

Another contribution to Melamine's phenomenal success was its design. As in the past, manufacturers turned to industrial designers to create wares that were functional yet aesthetically pleasing. These designers employed the 'form follows function' formula combined with plastic's organic quality to produce some truly unique and beautiful shapes. This period (1947-1955) could truly be called the Golden Age of Plastic Dinnerware. These wares can be identified by their heavy weight, quality production and colors. These early sets tried to mimic the solid color ceramic wares popular in the 1930s and 1940s (Fiesta, Bauer, Poppytrail) but using the color schemes of ceramic dinnerwares of the time (American Modern, Ballerina, Rhythm). Look for lots of greys, charcoal, burgundy, chartreuse, forest green and occasional Fiesta-like bright colors. The coupe shape and the rounded square shape, as well as the concepts of starter sets and stackable pieces, were all products of this incredible era.

By 1955, the plastic dinnerware scene moved into a new period. Pink (in all its shades), aqua, and turquoise exploded across the Melmac landscape. You will find many pieces of dinnerware in these colors,

Many 'starter' sets came boxed, like this Texasware dinnerware set.

often combined with other pastels. Around this time, a foil decal for Melamine was developed. These lithograph designs actually fused with the piece as it was produced (similar to underglaze decals on ceramic dishes). Now, Melmac could be decorated with any design imaginable. In an attempt to imitate and compete with china for home and restaurant use, the industry created endless floral, scenic, provincial, cutesy, and sometimes weird modernistic patterns for their dinnerwares. By the late 1950s there were almost three hundred patterns of Melamine dishes to choose from. Names like Regal, Windsor, Royalon, Debonaire and Imperial tried to give their products an elegant air worthy of fine china. Design-wise, styles were also changing. Coupe shape gave way to swoop shape. Thin was in, as was low-slung. This unfortunately gave rise to many cheaply made products advertised as "sleeker". Concurrent with this trend, manufacturers returned to one of plastics original purposes — to imitate other materials. Plastic ebonized wood bowls, imitation Raffiaware, acrylic serving pieces molded like cut glass, and a plethora of textural mimics followed. Take, for example, this item from a 1953 *Modern Plastics Magazine*: it shows a set of salad bowls, plates, and utensils, and says that "they are molded of sawdust and urea resin . . . and are richer looking than actual wood." Plastic was looking like anything but plastic! Even though it was estimated that in the early 1960s one in four families owned a set of plastic dishes, the era of plastic dinnerware was drawing to a close.

Unfortunately, Melamine dinnerware did not share the continued success that plastic kitchenware did. At the height of its popularity, ads for plastic dinnerware appeared in such publications as *The Saturday Evening Post, Good Housekeeping, House & Garden, New York Times Magazine, Farm Journal, House Beautiful*, and *Living*. Plastic dinnerware held so many promises. What happened? Some reasons can be cited for its demise in popularity:

1) Melamine simply could not replace china and ceramic dinnerware. Plastic's basic nature would always betray itself. Many critics feel Melamine's attempts to mimic or replace china was its downfall, as it turned away from its original use as a new and exciting medium.

2) Melamine did not live up to its much-touted reputation. Though practically unbreakable, Melamine was not indestructible as it was so often advertised. Repeated dishwashings, scrubbing with steel scouring pads, popping pieces in ovens to reheat food, and a multitude of other stress- and heat-related damages destroyed many Melamine pieces in a relatively short time.

3) Melamine dinnerware simply went out of fashion. The novelty lost its initial post-war sparkle by the 1960s. As people returned to more tried and true elements in style, pink and aqua gave way to avocado and autumn gold. Freeform dishes gave way to shapes that were 'traditional' (a big word in the early 1960s).

4) Competition. Plastic dinnerware wasn't able to maintain its ground against the wave of imported china, stoneware and porcelain that flooded the American market from Europe and finally the Far East. The release of Corelle glass dinnerware by Corning in 1970 eclipsed what was left of the already decimated Melamine market

A Melamine dinnerware set designed for children. Made by International Molded Plastics, Inc. in 1955. From *Modern Plastics Magazine*.

COLLECTING PLASTIC DINNERWARE

The first thing you need to understand is the definition of Melamine. Melamine is the name of the chemical used to make plastic dinnerware. "Melmac" is a brand name created for Melamine produced by the American Cyanamid Company. Other companies also produced Melamine but did not create a brand name for it. Don't be dismayed if you find a piece that does not say Melmac somewhere in its backstamp. It's still Melamine.

Sadly, very little Melamine dinnerware has survived in excellent condition. Misuse and lack of care sent much Melamine to an early grave as pet bowls under planters, or out to Dad's workshop. Finding sets in excellent condition is rare. All post-war Melamine, whether earlier or later in production, had a beautiful glossy finish and is quite appealing when found in this condition. Dulled finishes, scratches and heat damage are the most common culprits. The better the condition, the more valuable the investment. But don't pass up pieces with dulling or minor damages,. You can always use these pieces in your service and replace them with better quality pieces as you discover them.

Distinguishing old and newer Melamine is pretty easy. Outdated colors and styling should be your first clue. Check for backstamps. These are embossed on the underside of pieces. They usually (but not always) include the words "Melmac" or "Genuine Melamine," the name of the line or pattern or the manufacturer and location (e.g. Boontonware, made in Boonton, NJ).

All the post-war Melamine dinnerware covered in this book was made in the USA. If a piece is backstamped "Made in Korea," "Hong Kong," or "Taiwan," it is probably newer than the pieces discussed here. Be aware that some older Melamine pieces from Canada have been found and are worth buying. You will also find numerous unmarked pieces. Matching the shapes and colors will help identify these pieces. Remember, pieces with patterns date from 1955 and on.

'Collectible' stores are currently offering mostly the solid-colored wares. This parallels the American ceramic dinnerware trend; the solid-colored dinnerwares of the 1930s and 1940s were the first to be rediscovered and collected in the early 1970s. Occasional patterned pieces appear, but only the ones with modernistic 'fifties-type designs are currently desired. However, if the patterned Melmac appeals to you, go for it. This area of plastic dishes is vast and still considered esoteric by plastic dinnerware enthusiasts. You may not find whole sets. Expect to pay top prices at these shops as dealers become increasingly aware of plastic as collectible. Examine the pieces carefully — only items in good to excellent condition should command a high price.

happy the bride...
who gets beautiful, break-resistant
Melmac° dinnerware

So colorful, so smart in design, so lasting... that's dinnerware fashioned of MELMAC molding material. And it's so break-resistant, so difficult to chip or crack, that it will last for many anniversaries to come.

Ideal to give—or to get—MELMAC dinnerware comes in a variety of distinctive designs and colors. Use it for every day and every *special* day; it never loses its lustrous color and satin-smooth finish! And it washes so easily... and safely... by hand or machine.

Yes, happy the bride who gets your gift of MELMAC dinnerware —and why not treat yourself to a set? If not yet available at your favorite store, please write American Cyanamid Company, Plastics Department, 36 Rockefeller Plaza, New York 20, N. Y.

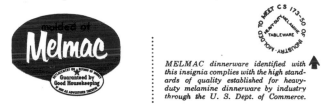

MELMAC dinnerware identified with this insignia complies with the high standards of quality established for heavy-duty melamine dinnerware by industry through the U. S. Dept. of Commerce.

Dinnerware Molded of MELMAC is sold under the following brand names:
BOONTONWARE, BROOKPARK, CODY'S, COLORFLYTE, DALLAS WARE, DESERT FLOWER, LIFETIME WARE, MALLO-WARE, MAPLEX, MELADUR, PROLON WARE, RESTRAWARE, TEXAS WARE

An advertisement for Melmac, featuring a happy bride and her friends. From *Good Housekeeping Magazine*, 1953.

All the great deals are still found at thrift stores and garage sales. Most second hand stores will toss the older plastic in with all plastics and the prices are usually reasonable. Garage sale prices are even better, as most people are more than glad to get rid of that "ugly old plastic." Again be forewarned; the majority of pieces you find will be in fair condition. You must decide if the particular piece is worth the price.

Sports cars, modern living...

and Boonton

Seems as though lots of folks have joined the trend toward leisurely living. Particularly the ones who appear in magazine, newspaper, and television advertisements. You can hardly find an ad, nowadays, that doesn't show a group of modern people smoking cigarettes, polishing the car, listening to music, or playing golf or tennis.

Not that we really object, mind you. Most of our customers spend pretty leisurely weekends themselves. That's because at Boonton we have a staff of people who do nothing but worry all day— about resins, specifications, cycle selec-

tion, mold temperatures, quality control, test runs, inspections, and all the other factors involved in a molding operation. As a result, our modern-type customers don't have to worry at all. They just leave each molding job with us and practically forget about it, until delivery time . . . on schedule, and exactly to specifications.

If you'd like, *you too* can join the trend toward smart, leisurely living. Just let Boonton take care of your compression and injection molding jobs. Then *you'll* be able to relax . . . and smoke cigarettes . . . and polish the car . . . and listen to music . . . and play golf or tennis . . .

BOONTON MOLDING CO.
BOONTON, NEW JERSEY

New York Metropolitan Area—Cortlandt 7-0003
Western New York Area—Alden 7134
Connecticut Area—Woodbine 1-2109 (Tuckahoe, N. Y.)
Philadelphia Area—Pioneer 3-0315

An advertisement for the Boonton Molding Company of Boonton, N.J., pushing its services more to the industry than to the consumer with this slightly whimsical, modernistic advertisement typical of the late 1950s.

An advertisement for Melmac featuring a Melmac plate "dropped 85 times from a height of 8 feet onto a hardwood surface." From *Good Housekeeping Magazine*, 1953.

invites compliments...defies breakage!
it's dinnerware
made of Melmac

Sturdy enough for everyday use . . . handsome enough for *best* . . . that's dinnerware made of MELMAC molding material. You'll love its smart good looks and its beautiful colors. Best of all, with MELMAC you can even ask the small fry to clear the table and help with the dishes. No more worries about breakage! Easily washed by hand or in automatic dish-washer!

Look for dinnerware made of MELMAC— it's probably at your favorite store in a variety of smart designs and colors. If not yet available, however, please write to American Cyanamid Company, Plastics Department, 36 Rockefeller Plaza, New York 20, N. Y.

"The two plates were dropped approximately 85 times from a height of 8½ ft. onto a hardwood surface," advises Ralph Bartholomew, Jr., photographer.

Typical of the wide variety of designs and colors available in MELMAC dinnerware.

WHAT ABOUT THAT COLD SUPPER?

The saving grace of sultry summer evenings — a cold supper, eaten casually outdoors, on the terrace. The public quickly realized that plastic dinnerware was perfect for this purpose. A *Modern Plastics Magazine* article from August 1936 said that "the plastics industry has been quick to anticipate the Twentieth Century housewife's needs and keep in step with her demands for efficiency, beauty, and modernity" — indoors or out! Lightweight dishes of Beetleware or other Ureas made in soft pastel blues and greens or alabaster white seemed to radiate a coolness that dispelled the hot, stickiness of the day. Beetleware's durability was also important for casual outdoor dining: ". . . the service is so durable that even the visiting bachelor, who offers to carry plates to and from the terrace, experiences no sinking feeling when a plate is accidentally dropped."

Hostesses were assured these cool dishes were quite appropriate for summertime bridge and buffet luncheons. They were also informed of other plastic items entering into the service of cool suppers: ". . . small plastic ash trays placed near attractive cigarette boxes trimmed with ivory or gay colored Ureas and the table itself may be covered with a beautifully inlaid surface of laminated plastic material in ebony black or even a pastel shade."

CLEANING AND CARE OF MELAMINE DINNERWARE

Alas, cracks, chips, burns and scratches can not be repaired. Dulled finishes can not be restored. However, much of the dirt and grime can be removed, and and pieces can be cleaned up for use. Dirty pieces purchased at thrift shops or sales should first be washed with a sponge or soft brush in warm to hot (not boiling) soapy water. I have found bleach helpful with some cleaning problems. A very small capful in your soapy water seems to restore some brightness, and disinfects pieces to be used. Always rinse pieces off thoroughly after cleaning with bleach. I have also found a quick (five-minute) soak in pure bleach will take out food stains imbedded in scratch marks. Rinse and wash immediately after bleach soaking and you will be surprised — the stain-imbedded scratches will seem to disappear. To remove those pesky glue-backed price labels from thrift stores and garage sales, use rubber cement thinner (also sold under the brand name Bestine). Using a cotton ball or rag, soak the label and then gently rub off. Remember, wash immediately in soapy water — *never* leave cleaners or chemicals on Melamine for *any* extended period of time.

The following is a short list of NEVERS in regard to cleaning and caring for your dishes:

NEVER clean Melamine in the dishwasher or soak in water overnight. Always clean by hand as soon as possible after use.

NEVER put Melamine pieces in or on a stove or grill to heat or cook food. (As far as microwave use, I have reheated foods on older Melamine for very short periods of time with no adverse affect. I would caution, however, against using older Melamine for microwave cooking.)

NEVER store food in Melamine dishes for any period of time

NEVER cut foods on Melamine. As mentioned, Melamine is highly scratchable. Scratching is one of the major causes of damage to dishes that have survived.

A traditional setting with Hemcoware dishes, made by Hemco Plastics. From *Modern Plastics Magazine*, 1948.

Chapter Five
Dinnerware

Dinnerware is defined as any item used at the kitchen or dining table to eat or serve food. This includes such basic items as plates, bowls, cups, saucers, and tumblers, as well as pitchers, shakers, butter dishes and serving pieces. Today, many plates, bowls, cups and saucers, creamers and sugar bowls are to be found, but the accessory pieces are harder to find. I have listed the dinnerware sets by the names of the lines. If one company made different lines, they will be listed together. The solid-colored lines will be shown first, followed by a section on patterned pieces. I will occasionally refer to a Melamine piece or line as as being heavy, medium or light in weight, as follows: *heavy* — thick, solid pieces that are noticeably heavy for plastic; *medium* — pieces of average weight, sometimes on the light side; *light* — thin and noticeably light.

Extensive research has turned up very little on the companies that produced plastic dinnerware. Many of these companies went out of business or were bought out with the demise of Melamine dinnerware's popularity. I have done my best to gather histories and information regarding the companies and their products. The results are, at best, very spotty. Remember, many of these companies were molders and manufacturers and made many plastic products — not just dinnerware. Hopefully, more research in the future will turn up valuable information.

ALLIED CHEMICAL

Allied Chemical's history in the plastic dinnerware field must have come much later because the pieces to be found reflect it. Medium to light-weight pieces marked "Allied Chemical" have been found in burnt orange, chocolate milk brown, cobalt blue, sage green, white and a dark shade of the ever-popular avocado. Allied Chemical is another backstamp you will always find at thrift stores. Pieces are easy to assemble into sets. Be forewarned: cups from Allied Chemical can be unmarked, and many unmarked pieces *not* from Allied Chemical seem to share the same shapes as Allied Chemical's shapes (the football-shaped serving bowl, for example.)

Allied Chemical items. (Back row) A round, patterned platter, 12"; oval platters, 12", orange and harvest gold. (Middle row) Oval serving bowls, 10', cobalt blue and orange; a bowl, 6", cobalt blue. (Front row, left to right) A harvest gold cup; a harvest gold bowl, 5"; a turquoise cup; a green bowl, 5.25", a green mug.

Allied Chemical items, including a platter, 13", and an oval serving bowl, 10", both in chocolate milk brown; an oval platter, 12", a cup and saucer, creamer, and 6" bowl, all in avocado; and a patterned plate, 6".

ARROWHEAD AND BROOKPARK

Created by designer Joan Luntz and manufactured by International Molded Products of Cleveland, OH, Brookpark and Arrowhead Everware share the popular squared shape. You will find more pieces marked Brookpark Modern Design than you will the older Arrowhead Everware backstamp. Brookpark appears to be a dead ringer for Franciscan's ceramic Tiempo line, which was created at the same time. Brookpark was first introduced in four colors: Chartreuse, Emerald, Burgundy and Pearl Grey. However, you will find Brookpark in numerous other colors, including black, an unusual and much-coveted color for dinnerware. Brookpark was another winner of the Good Design Award in the early 1950s. International Molded Products also made a line backstamped simply "Arrowhead." This line appears to be for institutional use as it features simple round table pieces and sectional serving trays.

A Melmac advertisement featuring the Brookpark pattern by International Molded Plastics, Inc. of Cleveland, Ohio. This is one of the most popular patterns among current collectors. From *Good Housekeeping Magazine*, 1954.

You're on the right track when it's tagged **MELMAC!**

BROOKPARK in Chartreuse and Emerald. Complete sets, open stock, serving pieces available.

BROOKPARK
unbelievably break-resistant
—of course it's genuine

Pearl Gray

Burgundy

melmac® dinnerware

In dramatic colors and interesting shapes —your modern design for dining, styled by Joan Luntz and made of MELMAC molding material. Use it for every meal, every day—MELMAC dinnerware is easy and safe to wash by hand or automatic dishwasher —safe, too, in children's hands.

Replacement Guaranteed

Free replacement, either where you bought it or from International Molded Plastics Inc., if any dish breaks, cracks or chips in normal household use within one year from date of purchase. Look for International's guarantee certificate — and look, too, for the tag that says MELMAC, sure sign you're getting the real thing!

MELMAC *is a registered trade-mark of American Cyanamid Company, N. Y. 20, N. Y., supplier of MELMAC Molding Compounds to manufacturers who fashion high quality dinnerware in a variety of designs and colors. Brookpark, shown here, is an exclusive design of International Molded Plastics, Inc., Cleveland 9, Ohio.*

dinnerware
melmac® *molded of*

Brookpark. (Back row) Cups and saucers in pink and turquoise; a 10" divided serving bowl in turquoise. (Front row) Plates, 9.5", 8", 6", in various colors; bowls, 4.5", in pink and black.

Brookpark. (Top row, left to right) A red cup; a 6" patterned plate; a 5" tall tab-handled bowl in blue. (Bottom row, left to right) Cups in burgundy and white; a bowl in jungle green (looks like a cup without a handle); a creamer in black.

Arrowhead items, including a yellow institutional tray, 10" x 14"; a 5" bowl in institutional green; a 9" white oval dish; a divided plate, 10", in chocolate milk brown; a 7" tan plate; a yellow 5.5" plate. Also pictured are a china restaurant creamer, salt and pepper shakers with plastic tops, a red Catalin-handled sugar dispenser, and Catalin-handled flatware.

AZTEC

The backstamp on this line pictures an Aztecan sun. The set is of medium weight, with low, flat plates and bowls like its western cousins Texasware and San Jacinto. So far, mostly pastels of blue, yellow, pink and white have been found. A sugar bowl of coppertone red and a round handled platter of mustard yellow are two exciting recent discoveries.

The Aztec line. (Back row) A divided serving bowl, 8", in beige-gray; a 10" yellow platter; a 7" white plate. (Middle row) Cups in chocolate and turquoise; a 6" turquoise bowl; a salmon and pink sugar and creamer set. (Front row) Cups and saucers in various colors.

BOONTON, BOONTON BELLE, AND BOONTONWARE

These lines were manufactured by the Boonton Molding Company, which started in Boonton, NJ around 1920. Their first Melamine dinnerware line was not offered to the public until 1947. The pieces marked "Boonton" are known as the "Commercial Line," and were introduced in 1948. Heavy, with stylish handles and knobs, many pieces have been found in pastel green, yellow, grey, forest green, burgundy, pink, and turquoise. "Boonton Belle" was created by designer Belle Kogan, and pieces were backstamped "Boontonware." Its colors, exotic rounded square shape, and light weight give it a more 'tasteful' look than the Commercial Line and date it to the mid-1950s. Some speckled pieces have been found in the Belle line. Belle pieces are harder to find than the more ubiquitous Commercial Line. You will also find pieces with the Boontonware backstamp that are not part of the Belle line. Though their colors are similar to the Belle line, their shape is round and the design is simpler. In addition, Boonton offered a Somerset Line and a Patrician Line.

Boonton continued producing Melamine dinnerware until 1977, when it sold its molds to a company called "English and English." Boonton officially ceased to exist by 1981. However, English and English is going strong, producing Boontonware lines at this time. Though Boonton was using some of its original molds for sets manufactured in the 1970s, a current brochure from English and English shows no older molds in the two lines being currently offered: an all-white line with table ware and serving pieces, and a more interesting line called the Confetti Line, white with blue-gray speckling. Boonton has produced some of the best quality Melamine dinnerware available, and it is good news that someone is continuing to produce these wares. Maybe some day they will re-issue the original lines. If you are interested in purchasing contemporary Boontonware you can write to English and English Inc. at P.O. Box 23, Bloomingdale, NJ 07403 for a brochure.

A Melmac advertisement for Boontonware. From *Good Housekeeping Magazine*, 1952.

From the Boonton line, a sugar and creamer in institutional green; a 10" yellow plate; a 9" beige plate, an 8" turquoise plate; a 7" yellow plate; wide-rimmed bowls, 6.75", in burgundy and institutional green; round-lipped bowls, 5.25' and 6", in jungle green and light blue.

From the Boonton line. (Back row) A yellow platter, 9.5", and a pink platter, 12". (Front row) A turquoise butter dish, and two 8" divided serving bowls in pink and institutional green.

From the Boonton line, a group of cups in the most popular colors available: burgundy, ivory, yellow, jungle green, pink, and turquoise.

From the Boonton Belle line. (Back row, left to right) A salt and pepper set in turquoise; a cup and saucer in turquoise; a gray dinner plate, 10". (Front row, left to right) 10", 8", and 6" plates in pink, turquoise with white speckles, and yellow; 5.5" bowls in pink and white; a 6" jungle green plate.

From the Boonton Belle line, a 14" serving bowl in dark charcoal, and a color selection of 6" plates: turquoise, white, pink, peach with white speckles, and yellow. *Serving bowl courtesy of David King.*

Miscellaneous Boontonware, including a 7" pink serving bowl; pink and yellow 5" bowls; cup and saucer sets in white/turquoise and pink/turquoise; a 6" yellow plate; a pink sugar bowl with a missing lid. The large pink plate in the back is from the Boonton Belle line; note the difference in styling.

Miscellaneous Boontonware. (Top row, left to right) A pink creamer; a pink saucer marked "Somerset," made in Boonton, NJ; a 4" turquoise bowl. (Bottom) Cups with different styling from those previously shown, in burgundy, charcoal, and gray.

BRANCHELL, COLORFLYTE, ROYALE

These heavyweight dinnerware lines offered by the Branchell Company of St. Louis, MO are particularly appealing due to their mottled appearance. So far, only the three lines have been found. The general line, backstamped "Branchell," comes in a variety of colors and shapes; Colorflyte and Royale are practically the same, sharing some of the general line's shapes and a similar color palette. Plates, cups, saucers and serving pieces have been most commonly found. All three lines are credited as 'designer dinnerware' by Kay LaMoyne, and were proclaimed as "a Melmac product of finest quality."

COLOR-FLYTE, available in mixed or matched colors—open stock, serving pieces. Write for free descriptive folder.

A Melmac advertisement featuring Colorflyte by the Branchell Company. From *The American Home*, 1954.

All set...on the prettiest tables yet!

Color-FLYTE
genuine break-resistant melmac® dinnerware

Beautiful, isn't it? Best of all, it's made of Melmac molding material to shrug off accidents . . . whisk safely through automatic washers . . . survive even kiddie-capers! See it! Get it! You'll love it!

Replacement Guaranteed by the Branchell Co., if any dish breaks, cracks or chips in normal household use within one year from date of purchase. Look for the Branchell guarantee certificate—and for the tag that says Melmac, sure sign that you're getting the real thing!

Melmac is a registered trade-mark of American Cyanamid Company, N.Y. 20, N.Y., supplier of Melmac Molding Compounds to manufacturers who fashion high-quality dinnerware in a variety of designs and colors. COLOR-FLYTE, shown here, is an exclusive design made by The Branchell Company, St. Louis 10, Missouri.

From Branchell. (Left to right) A copper-colored serving bowl, 10.5"; a salmon red serving platter, 14"; a light blue serving platter, 12"; a blue ashtray marked "Designed by Kay La Moyne."

Debonaire

This 'sophisticated' set is a real find — if you can find it! Most pieces are speckled, and the lone patterned cup is interesting. The backstamp gives no information except the name Debonaire.

From the Debonaire line, a 10" pink serving bowl, a 7.5" low yellow bowl, an 8" yellow bowl, a yellow creamer and sugar, and a turquoise cup with a white saucer. The saucer is decorated with stylized leaves.

Durawear/Capac

This line by California Molded Products, Inc. of Santa Paula, CA, is another fun and futuristic-looking line. Though there are two different backstamps (Durawear and Capac), it is obvious they are the same line, or were made concurrently by the same company. Real 1950s styling and colors are a big plus. Though they date from the late 1950s, these medium-weight pieces are sturdy and appealing. Evidence suggests that Durawear may have been created by a prestigious designer of the times.

From the Durawear line, three 10" serving bowls in pink, yellow, and salmon red.

From the Durawear line, 10" plates in pink and dark charcoal, and cups and saucers in blue and pink.

From the Durawear line. (Back row) Bowls, 7.5", in chocolate milk brown and turquoise; a 5.5" white bowl; a cup and saucer set in white and salmon. (Front) A color selection of speckled 7" plates, in ivory, turquoise, mustard yellow, and salmon red.

From the Durawear line, a 14" platter in salmon red with dark red speckles; creamers and sugars in turquoise, pink, chocolate milk brown, and salmon red.

FOSTORIA

The set shown here was made by the Fostoria Glass Company, and the backstamp features the same logo the company's glass factory used. This plastic pattern is heavy, durable and quite appealing. It is two-toned, a style popular in the later 1950s. The plates and saucers have aqua bottoms with white tops adorned with pale aqua, stylized, daisy-like flowers as shown. The serving bowl and shaker are in solid aqua. It is quite an elegant set for Melamine.

From Fostoria, a 10" dinner plate, an 11.5" platter, a shaker, a cup and saucer, a 10.5" divided serving bowl, and a 6" plate.

The pattern on Fostoria's platter can be seen in this close-up shot.

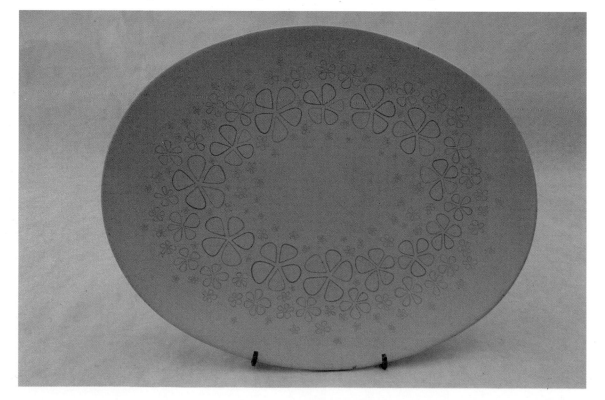

GOTHAMWARE

So far, no information is available about pieces with this backstamp, but some backstamps also read "Made in USA" or "Made in USA and Canada." All items found so far are made of Polystyrene. Some pieces are speckled with grey and look like Melamine dinnerware, while other pieces are in bright, solid colors, and seem to have been made for picnic purposes. The small blue bowls have the original labels which read "Cymac — a Methylstyrene Product."

From the Gothamware line, two 6" utility bowls of Polyethylene, blue and yellow; mugs in various colors; a divided plate in yellow; a different divided plate in red, blue, gray, and chartreuse.

From the Gothamware line, a selection of speckled tableware including 9" plates in blue, pink, and yellow, a 6" plate in yellow, and a 4.5" bowl in blue; solid-colored tableware including leaf-shaped coasters in salmon red, deep blue, and gray. Also pictured (not Gothamware) are an acrylic spangled pitcher, Polyethylene tumblers, and a paper salt and pepper set with plastic tops.

Gothamware cups and saucers, and a 6" yellow plate.

HARMONY HOUSE: TALK OF THE TOWN, CATALINA, AND TODAY

Sears and Roebuck had many items made for their stores and catalogs under the brand name Harmony House. In the 1953 catalog, Sears introduced their first Harmony House Melmac line called "Talk of the Town," in the square shape. The set came in four colors — Dawn Grey, Mint Green, Victorian Red and Chartreuse. The pieces available included a 10" dinner plate, a 7" salad plate, teacup and saucer, soup or cereal bowl, sauce dish, creamer and sugar bowl, open vegetable dish and a 13" platter. They were advertised as being "so durable, so beautiful and its only $12.95 for a 16 piece set"

The lightweight quality of plastic dinnerware was a boon to Sears, at least in regard to mail order shipping costs. In the spring of 1954, Harmony House released the Today line. This line was less stylized and featured plain round shapes. The original four colors were Spice Beige, Aquamarine, Sunshine Yellow, and Dawn Grey. A 16-piece set cost $11.95 and included all the pieces above plus a 6" bread and butter plate. In 1955, Catalina was introduced as "a popular new coupe shape" in four colors: Inca Gold, Bronze Green, Malibu Coral, and Spice Beige. At the same time Talk of the Town was renamed "New Talk of the Town," and four pastel colors were added: Frosty Pink, Clay Beige, Medium Federal Gold, and Medium Sage Green.

In 1957, Sears introduced its first products in the New Decorated Melmac line: the Frolic, Province, and Crocus patterns on square-shaped dishware, and the Patio Rose and Autumn Leaves patterns on dishes in the Catalina (coupe) shape. Also introduced that year was a line called Catalina Translucent, which came in four solid colors: Ming Blue, Light Malibu Coral, Parchment Beige, and Light Federal Gold. These pieces are quite appealing owing to their translucent quality.

Over the next few years, patterned Melmac proliferated. By 1960 Sears was retaining a certain Count Sigvard Bernadotte to design dinnerware lines for Harmony House. The Count's first creation was called Golden Spears (stalks of wheat) and was advertised as "Grace and beauty in tall flared shapes and a high footed, off-the-table look." Count Bernadotte designed other lines for Sears. Sadly, the solid-colored lines were slowly fading — only two new solid lines had appeared: Windsor, advertised in 1959 as a "medium weight Melmac" in Ming Blue, Shell Pink, Parchment Ivory, and Sunshine Yellow, and 1961's Easy Livin' in Pink, Turquoise, Ivory, and Sunshine Yellow. By this time, however, Catalina, Today, and Catalina Translucent had all been discontinued, and were removed from Talk of the Town (now called "Talk of the Town Savoy").

You will readily find Talk of the Town pieces in your search, obviously due to their longer production runs. They are really neat pieces and good medium-weight quality. The other lines you will not find so easily.

From *Talk of the Town*'s Harmony House line. (Back row, left to right) A 10" divided serving bowl in green, and a 10" dinner plate in pink. (Front row, left to right) Cup and saucer sets in pink, and a sugar and creamer in chartreuse.

From *Talk of the Town*'s Harmony House line. (Back row) A 10"
divided serving bowl in yellow. (Front row, left to right) 10" dinner
plates in chartreuse, pink, and green. Shown with a salt and pepper set made by Hazel Atlas Glass Co.

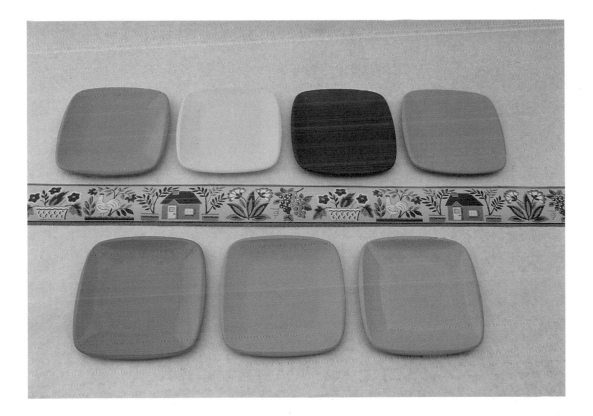

From *Talk of the Town*'s Harmony House line, a color selection of
6" plates: chartreuse, yellow, mint green, pink, green, gray, and
beige.

From the Harmony House line. (Top row) Two 12" platters in pink and white; a creamer in mustard yellow, marked "Catalina"; a sugar with a white body and a blue top, marked "Catalina"; a turquoise mug marked "Today"; a pink creamer marked "Catalina." (Bottom row) A 5.5" turquoise bowl marked "Avalon"; a saucer in mustard yellow, marked "Catalina"; a 5.5" bowl in chocolate milk brown, marked "Catalina"; a 5" yellow bowl marked "Today"; a 5" bowl in mustard yellow, marked "Catalina." *White and blue sugar courtesy of David King.*

HEMCOWARE

Hemco Plastics was one of the very early manufacturers of plastic housewares and dinnerwares. As early as 1938, Hemco was offering "informal dining sets" in Urea, continuing until after the war. Hemcoware comes in an array of colors and shapes. Early pieces sport plain colors such as pale green, ivory, and light blue. Hemco also produced numerous kitchenware items in other plastics. Keep an eye out for their early institutional ware, particularly their wartime and airline pieces.

From Hemco Plastics' Newport line: 10" dinner plates in turquoise and yellow; a cup and saucer set in jungle green and burgundy; a 5" bowl in jungle green; a 7" bowl in gray; an 8" plate in jungle green. *Burgundy saucer courtesy of David King.*

From Hemco Plastics. (Top row) Three mugs in light blue, chocolate milk brown (both 3" tall), and yellow (2.75" tall); a 2.5" tall two-handled mug in orange, marked "Genuine Beetle." (Bottom row, left to right) A 5" yellow bowl; a two-handled mug in yellow; a 2.75" tall red tumbler; a 3.75" tall green tumbler marked "Airline."

Hemco Plastics' two-handled orange mug has this paper label on its base. "Beetle" was the name for their Urea products. Hemco pieces found today might include some Urea, which the company used for numerous table pieces.

From Hemco Plastics' Hemcoware line. (Back row) Divided plates, 10.5", in institutional green and pale blue. (Middle row) An oval serving bowl, 10", yellow. (Front row) Divided plates, 9.5", in yellow and brown; a divided bowl, 7.5", in green. Also pictured is a wooden potholder hanger decorated with a wooden rooster.

Hemcoware. (Back row, left to right) A yellow plate, 9.5", and a green plate, 8". (Middle row) Plates, 7", in yellow, white, institutional green, and peach. (Front row) A Polystyrene coaster set in yellow, green, red, and blue. Also pictured is a coffee canister in turquoise Polystyrene, not by Hemco.

HOLIDAY

This medium-weight line of Melamine dinnerware, made by the Kenro Company in Fredonia, WI, is ultra-modern in styling. It has flared shapes and is usually speckled with red. Regrettably, very few pieces of this appealing pattern have surfaced. The small tumbler is backstamped "Capitol Airways," most likely an airline from the mid-1950s.

From the Holiday line, two 10" plates in turquoise and yellow; a 5" tumbler in turquoise; a 3.5" tumbler in ivory; a pink creamer; a 6" bowl in institutional green; a red cup; a 5" low bowl, in yellow. All pieces except the creamer and the green bowl have dark red speckles.

IMPERIAL WARE

Almost all that is known about this line is that its backstamp reads "Imperial Ware," with no indication of what company produced it or where. The line does include many speckled pieces in various color combinations. This is good, medium-weight ware with simple designs, and it should spark collectors' interest if more of it surfaces.

From the Imperial Ware line, three 10" dinner plates – one in pink, two in white with speckles; a 6" plate, white with speckles; a 6" bowl, white with speckles; two cup and saucer sets, turquoise and yellow cups with white speckled saucers. *All white speckled pieces courtesy of David King.*

From the Imperial Ware line, an 8.5" serving bowl in pink; a creamer in jungle green; saucers in pink, burgundy, institutional green, and gray; 6" plates in yellow and jungle green; 6" bowls in various colors.

LIFETIMEWARE

In 1915, The Watertown Manufacturing Co. was formed in Watertown, CT. At first, they produced buttons and electrical parts from the resin compounds available at that time. By 1930, they had a big new factory and were producing mostly plastic items. During this period they produced a vast list of articles using a custom plastic material called Neillite. Neillite was also known under the names Tenite and Lumarith. Electrical housings were still the main use of Neillite, but many other items (including fuses, switches, and appliance housings and bodies) found ready use for this material.

After World War II, the company manufactured one of its most successful products, a line of plastic dinnerware. The design for these dishes, the work of a local artist named Jon Hedu, was extremely advanced for its time. The company's concept of colorful, modern dishes that were attractive and unbreakable was considered revolutionary, and the designs received many awards.

A large part of the firm's plastic dinnerware business was the U.S. government's purchase of plastic dinnerware to be used on navy ships. The Watertown Lifetimeware line became well-known the world over. By 1948, Lifetimeware became one of the first Melamine dinnerware sets to be available to the public.

A Melmac advertisement featuring Lifetimeware by Watertown Mfg. Company. From *Good Housekeeping Magazine*, 1953.

*Lifetime Ware in Granada Green and Canyon Yellow.
Complete sets, open stock, serving pieces available.*

Lifetime

beauty that needs no pampering—it's genuine break-resistant melmac dinnerware!

Bermuda Coral   Sahara Sand   Caribbean Blue   Palisades Grey   Chartreuse   Cocoa

The set that gives you years of care-freedom—Lifetime by Watertown! The beautiful every-meal dinnerware that needs no care in handling, it's so remarkably break resistant; no extra care in washing, either, it's so easy and safe to wash by hand or automatic dishwasher. Yet, Lifetime Ware is so *care-fully* designed, so functional, so lovely... and it's

GUARANTEED AGAINST BREAKAGE!

Replacement by the maker, Watertown Manufacturing Co., guaranteed if any piece breaks, cracks or chips within a year in normal household use. Look for their guarantee—and for the tag that says Melmac, sure sign you're getting the real thing!

MELMAC *is a registered trade-mark of American Cyanamid Company, N. Y. 20, N. Y., supplier of* MELMAC *Molding Compounds to manufacturers who fashion high quality dinnerware in a variety of designs and colors. Lifetime Ware, shown here, is an exclusive design of Watertown Manufacturing Co., Watertown, Conn.*

From the Lifetimeware line, a color selection of cups: light blue, beige, black, red, yellow, and pink.

From the Lifetimeware line. (Back row, left to right) A 10" oval serving bowl in gray; a creamer and sugar in chartreuse and light brown. (Front row, left to right) A turquoise butter dish; a 10" plate in chartreuse; an 8" plate in light brown; a 5" tab-handled bowl in yellow; a yellow cup and saucer; an 8" plate in chartreuse; a 7" bowl in blue.

From the Lifetimeware Translucent line. (Back row, left to right) A 14" oval serving platter in pale green, and a pale green water pitcher. (Front row, left to right) All in pale pink, a creamer and sugar set; a 10" divided serving bowl; 5" bowls; and a cup and saucer. Also pictured are a Vernon Kilns Tickled Pink ceramic fruit bowl, and a ceramic flamingo figurine on a Polyethylene doily.

MALLO-WARE

Mallo-ware, made by P.R. Mallory Plastics Inc. in Chicago, is attractive because of its color palette, the number of pieces, and its sturdy, simple design. Each piece is identified on the back with a stock number. Some pieces are backstamped "Mallo-Belle." This is possibly a later lightweight restyling of the original Mallo-ware line. The tumblers shown are an unusual find. The P.R. Mallory Plastics company also created the well-known Raffiaware line featured in the kitchenware section.

From the Mallo-ware line. (Back row) A 14" platter in yellow. (Front row) A chartreuse sugar bowl; a white creamer; a 9" divided serving bowl in orange-red; a 5" bowl in beige; a 6" plate in institutional green.

From the Mallo-ware line. (Back row) Three 9.5" plates in orange-red, institutional green, and pink. (Front row) four 4.5" tumblers in turquoise, orange-red, white, and yellow; cups and saucers in pink and turquoise.

From the Mallo-ware line, a color selection of 5" tab-handled bowls: pink, white, jungle green, orange-red, brown, yellow, and turquoise.

162

MARCREST AND MARCREST PASTELS

The Western Stoneware Company made a popular line of dark brown, heavy stoneware for the Marcrest Company in the 1950s. A series of plastic dishes bears a Marcrest backstamp that is very similar to the backstamp on this stoneware, so presumably they were made by the same company. Brown must have been their favorite color, as many of the Melamine pieces can be found in chocolate milk brown. There is (pictured) a different line marked "Marcrest" that comes in appropriately soothing 1950s pastels. Not a high quality product, this is another example of the ceramic industry's attempt to cash in on the Melamine phenomenon of the 1950s. Both lines are medium-weight to lightweight.

From the Marcrest line, a creamer in chocolate milk brown; a white sugar, a 6" turquoise bowl, 5.25" and 5" bowls in chocolate milk brown, an 11.5" divided serving bowl in chocolate milk brown, and a cup and saucer set in turquoise and white. Also shown is a McCoy ceramic bud vase.

The Marcrest Pastel set comes in four colors: pink, blue, yellow, and white. Shown here are a 14" platter, 10" dinner plates, 7" plates, and 6" and 5" bowls.

From the Marcrest Pastel set, a 9.5" divided serving bowl, a creamer and sugar, and cup and saucer sets.

MIRAMAR AND MIRAMAR LAGUNA

These later lines were made by Miramar of California in Los Angeles, California, and they are typically less sturdy than older lines. Turquoise and white pieces are most often found. You will find many patterned pieces with the Miramar backstamp. Notice the 'pixie-ear' handles on cups and sugar bowls! Some pieces are backstamped "Weber Plastics L.A. Calif."

From the Miramar line, sugar bowls in avocado and turquoise; cups and saucers in white/pink and white/turquoise combinations; an 8" serving bowl in chocolate milk brown. Also pictured is a freeform ceramic candy dish with gold starbursts.

MONTEREY MELMAC

The "Watertown" backstamp on pieces in this line suggests that it was made by Watertown Mfg. Co., the manufacturers of Lifetimeware. Pleasing, solid colors and good medium weight make this a delightful pattern.

From the Monterey Melmac line. (Back row, left to right) Bowls, 5", in pale blue and yellow; a 10" plate in brick red; a 6" plate in gray; an 11" platter in yellow. The right half of the platter has been cleaned with a five-minute soak in bleach. (Front row) Cups and saucers in brick red, gray, and light blue.

We May Be Gone - But Our Dishes Survived

What would happen to plastics if the locations of their use were subjected to atomic attack? To answer this question, Project 31.5 was set up in Yucca Flats, Nev., as part of 'Operation Cue' in the spring of 1955. This was a cooperative venture conducted under the auspices of the Society of The Plastics Industry, Inc. The purpose of Project 31.5 was to test the thermal ignition of plastic materials under atomic blast circumstances.

To conduct this experiment an atomic device labeled "Apple 2" was placed atop a 500 ft. high iron tower. The 680 plastic units involved in the project were installed at three stations located at 6000 ft., 7660 ft., and 8700 ft. from Ground Zero (the center point of the explosion). Plastic samples were cut into 12" x 12" squares, mounted and hung 'clothes-line-style' on wire lines. In addition, plastic items like Melamine dinnerware and kitchen and houseware items made of various plastics were hung up to be tested.

The degree to which these samples were influenced by the thermal effect of the blast depended upon: 1) the type of material; 2) the thickness of material; 3) color; and 4) distance from Ground Zero. The results are as follows:

Vinyl squares — at Station 1 (the closest point of measurement) the material melted completely; at Station 3 it fused.

Teflon squares — remained intact.

Polyethylene squares — thicker examples remained intact at Station 1.

Polystyrene squares — experienced some melting at Station 1; thicker pieces remained intact at other locations.

Acrylic squares — thicker pieces remained intact.

Melamine dinnerware — at all three stations, remained unaffected and undamaged.

One final note: color (or probably infrared reflectance) was an important factor in these tests. White and transparent samples exhibited less distortion, charring, and melting than did the darker colors. Dark-colored samples were almost always affected, and black faired the worst.

Oneida, Oneida Premier, and Oneida Deluxe

China and glass companies tried producing plastic lines in the late 1950s, though they met with little success — but a silver company? Research has turned up very little as to the origin and history of the plastic lines produced by Oneida Ltd. in Oneida, New York, and it is uncertain whether this company is related to the famous Oneida Silver Company. These plastic pieces are not terribly appealing or well-made, but expect to find them in early to mid-1960s colors — golden harvest, white, avocado, and the ever-present 'chocolate milk' brown. Many pieces are unmarked. The design of the gravy boat and underplate is interesting.

From the Oneida line. (Back row, left to rigth) A 6" peach-colored plate; a 4" turquoise bowl; a 10" beige serving bowl; a 3.5" tall red and white creamer; a 7.5" plate in cobalt blue. (Front row, left to right) A cup and saucer set in mustard yellow; a 6" bowl in avocado; a gravy boat with underplate in mustard yellow.

Prolon, Prolon Ware, Florence, Leonora, and Beverly

The Prophylactic Brush Company of Florence, Massachussetts started in 1866 as the Florence Manufacturing Company. The company made hairbrushes, buttons, pistol cases and other products out of an early natural resin compound. In 1884, they started making toothbrushes out of cattle shinbones. At the turn of the century, the company found Celluloid very adaptable and started producing toothbrushes and hairbrushes from this material. By 1940, ProBrush (as it was and is now called) was the first company to make brushes from methyl methacrylate (acrylic). Later a lower cost line of hairbrushes, toothbrushes, combs, etc. were produced under the highly successful Jewelite line. In 1942, the Prolon division of ProBrush was formed, and the first plastic dinnerwares by this company were produced shortly after. By 1952, Prolon's General Line was in full production. Today, ProBrush is one of the largest toothbrush manufacturers in the world.

Pieces marked "Prolon" or "Prolon Ware" I call the General Line. Prolon pieces are older and heavier. Prolon Ware offers thinner pieces in mid-1950s colors. Both lines are commonly found. Pieces in the Beverly and Leonora patterns rarely surface, and should be considered rare. The Florence line, created by designer Irving Harper, was inspired by Japanese laquerware. This much sought after 'designer pattern' helped boost Melamine's reputation. It won the *House Beautiful* Classic Award when it was introduced in 1953. You will find patterned as well as solid pieces in this line.

From the Prolon line. (Left to right) A burgundy sugar bowl, a yellow creamer, and a yellow sugar bowl.

From the Prolon line, all in turquoise. (Back row, left to right) A sugar and creamer, a 10" plate, and an 11" oval serving bowl. (Front row, left to right) A cup and saucer, a 10" plate, a 6" plate, a 5.5" bowl, a 4.5" rimless bowl. Also pictured is a Steri-lite salt and pepper set.

From Prolon's Beverly line and (if so marked) from Leonora: a 14" yellow platter; a 10" turquoise plate; a pink creamer; a 6" pink plate; a 10" pink plate marked "Leonora." *Plate marked "Leonora" courtesy of David King.*

From the Prolon line. (Top row) Dinner plates, 10", in burgundy, pink and white; plates, 6.5", in jungle green, beige, and red; cups and saucers in combinations of yellow and red, red and pink, and turquoise and yellow.

From the Prolon line. (Back row) A 14" platter, in yellow, marked "Grant Crest by Prolon"; a 10" divided plate in institutional green; a 14" platter in white. (Middle row) A 4.75" yellow bowl marked "Grant Crest by Prolon"; 5.25" bowls in chocolate, mottled, turquoise, and pink. The mottled bowl is styled differently and has an older backstamp. (Front row) A white creamer.

From Prolon's Florence line, 10" dinner plates in red and blue, and cups and saucers in red, blue, and white.

From Prolon's Florence line, a 14" beige platter; a 9.5" divided serving bowl in turquoise; a 9.5" serving bowl in mustard yellow; a 6" plate in beige; a 12" platter in turquoise.

RIVIERA-WARE

Another mystery line. A heavy-weight line with simple shapes and pastel colors, it resembles a Prolon-type ware. This probably dates it from the early 1950s. Plates, cups, and saucers are most readily found, though they may not be in good shape.

Riviera Ware: Dinner plates, 10", yellow, institutional green and tan; Bowls, 5.5", tan and blue; Cup and saucer sets in various colors.

ROYALON AND ROYMAC

Royalon Inc. of Chicago made these obviously late-1950s/early-1960s lines, which must have been popular; they certainly are different. The pieces found so far center around combinations of bright cherry red and white. Cups can be all white, or white inside with red outside. Saucers, plates, and bowls are white with red underneath, all red, or all white. However, a pastel lavender serving bowl has been found!

From the Royalon line. (Back row) Divided serving bowls, 10", in lavender and red; a saucer, showing the red bottom. (Front row) Cup and saucer sets, in two-tone red and white, and turquoise and pink; a serving bowl, 9", in pale pink.

RESIDENTIAL AND FLAIR

The Northern Industrial Chemical Company of Boston, Massachussetts has been in the plastics industry since before World War II. During the war, Northern did an enormous share of plastics manufacturing for machine parts, weaponry, and dinnerware for the armed forces. They were also a major supplier of the first plastic dinnerware to early airline companies. Original pieces were made of Beetleware (Urea), but the company moved on to Melamine after the war as airline business increased.

Northern was so confident that in 1947 the company advertised Northern Air-Ware for domestic use. A six-piece place setting in ivory Melamine sold for $1.95 and included two plates, a bowl, a cup, a saucer, and a tumbler. Advertising copy tells us this is a ". . . durable, lightweight airline type of Individual Place Setting. . . for trailer, camp, factory cafeteria and home breakfast nook." The same year, Northern introduced a tableware line for hotels and restaurants. The line came in three colors and had specifically designed 'feet' for easy stacking.

Northern's leap into the spotlight came in 1953, when they released their Residential line of Melamine dinnerware. This line was created for the company by Russel Wright, the famous industrial designer. Many claim that Wright's entry into the Melamine dinnerware field was the turning point in the public's acceptance of plastic dishes. Wright was already a popular household name due to his furniture, chrome pieces, and ceramic dinnerware lines.

His designs for plastic dinnerware were refined, heavy, richly colored, and sold in the millions. Another line designed by Wright, Flair, was also successful when released in 1959. Because of Wright's name association and his superior design skills, pieces from these sets are the most highly sought after on the market today. For some reason, not much has surfaced, at least on the West Coast. (Wright designed two more plastic dinnerware lines, Home Decorator and Idealware, for other companies.)

Idealware by Russel Wright, 1958. From *Good Housekeeping Magazine*. Wright's sincere effort to create dinnerware using Polyethylene did not meet with the overwhelming success his earlier wares had.

Residential and Flair pieces (Residential unless marked "Flair"). Back row: Plates, 7.5", tan, 10", pink marked Flair. Front row: Divided serving bowl, 11", yellow; Cups, mottled gray and mottled blue. *Blue cup courtesy of David King.*

SPAULDING AND SPAULDINGWARE

There appear to be two different phases in the production of this highly colorful line, made by the American Plastics Corporation in Chicago, Illinois. The pieces marked "Spaulding" are noticeably heavy. Pieces marked "Spauldingware" are lighter, though the designs are the same as the heavier Spaulding pieces. There are even some odd pieces that feel as if they might be made of Polystyrene.

From the Spaulding line: (Top row) A creamer and sugar, mustard yellow; a 12" platter, yellow; a 10" serving bowl, yellow. (Bottom row, left to right) A creamer and sugar in two different styles, turquoise and pink; Cups, yellow and pink. *Yellow platter and bowl courtesy of David King.*

From the Spaulding line, an 11" divided serving bowl in mustard yellow, 6" plates and cup and saucer sets in red, mustard yellow, and gray.

From the Spaulding line, a color selection of 10" dinner plates: yellow, chocolate, pink, gray, and turquoise. Also shown is a Spaulding tab-handled bowl, 4.5", in red.

STETSON AND SUN VALLEY

Other entries into the pastel craze of the mid-1950s are these two apparently identical patterns from Stetson Chemicals in Lincoln, Illinois. Pieces backstamped "Stetson" are more common than those stamped "Sun Valley." The colors offered were the standard pink, aqua, yellow, and white. Still, a good number of pieces of Stetson have surfaced — I'd like to see how easy it will be to collect this medium-weight and predictable line.

From the Stetson line, a 12" platter in chocolate milk brown; cup and saucers sets in pink/white and blue/pink combinations; a pink gravy boat; a pink sugar; a salmon red mug; a pink butter dish.

From the Stetson line. (Back row) A 14" pink platter; a 7" turquoise plate; a turquoise sugar and creamer. (Front row) 9" plates in turquoise and yellow; 4.75" and 5.25" bowls in chocolate milk brown and pink.

From the Stetson line, a selection of pink pieces. (Back row, left to right) A 9" divided serving bowl, and a 14" platter. (Front row, left to right) A cup and saucer, creamer and sugar, 7" plates, and 5.5" bowls.

Texas Ware, Dallas Ware, San Jacinto, and Rio Vista

I am told that the manufacturer of these lines, the Plastics Mfg. Company of Dallas, Texas, still exists and is still producing Texasware. This may explain why any given trip to a thrift shop will turn up a few neglected pieces of Texasware. Tons of this stuff must have been (and is being) made. I have seen heavyweight plates in every color from 1940s institutional green to hot orange, lightweight electric blue and lime pieces, and patterned pieces with very 1970s and 1980s designs. Texasware is a field in itself to collect, and those shown here are just a sample.

Dallasware seems to be institutional ware, not very colorful or interesting. Still, assembling basic table settings is relatively easy. The San Jacinto line was available in six solid colors and four designs, circa 1953. The mottled brown and white pieces are called San Jacinto Contemporary. The Rio Vista line is not often seen. It sports solid colors and two parallel wavy lines near the rim of plates. Texasware and Dallasware are looked down upon because of the ubiquity of the cheaper wares, but it would be nice to see more products from this company emerge — particularly in the San Jacinto and Rio Vista lines.

A Texas Ware advertisement featuring San Jacinto pattern in solid colors. Made of Plaskon Melamine. From *Good Housekeeping Magazine*, 1954.

San Jacinto line / Rio Vista line: Plates, 7", turquoise and mustard yellow; low bowl, 6.25", pink, all marked Rio Vista; Plate, 7", white, saucers, brown and turquoise, all marked San Jacinto.

Dallas Ware Line. (Back row, left to right) A 9.5" divided plate in institutional green; 3" mugs with grooves for stacking, in tan and blue; a 10" plate in cobalt blue. (Front row, left to right) A 9" oval serving dish in white; a 7" low bowl, in tan; a 5.5" bowl in institutional green.

From the Texas Ware line. (Back row) A 14" yellow platter, two 10" plates in peach and blue; and lime green and cobalt blue saucers. (Front row) A turquoise cup and patterned saucer; a 6" yellow plate; a 5" turquoise bowl; an 8" orange-red mixing bowl; an 8.5" turquoise serving bowl; a 10" white plate; a green mug; and a turquoise sugar bowl.

From the Texas Ware line, two 12" platters, a sugar bowl, and an 8.5" divided serving bowl (all white with brown speckles); a 12" platter, a 6.5" plate, a 5" bowl, a cup and saucer, and a butter dish (all in chocolate milk brown).

WINDSOR MELMAC

Like its color cousin Stetson, Windsor runs the gamut of pastel shades on its medium-weight, durable wares. Many basic pieces have been found. The round serving bowl with an S-shaped division is the same shape used for the Stetson set. Still, it is unknown whether the two lines are related.

From the Windsor line, a 9" divided serving bowl in white; a yellow cup and saucer set; pink and turquoise cups; a turquoise sugar (lid missing) and creamer; and a pink sugar.

MISCELLANEOUS SOLID-COLORED DINNERWARE: WITH BACKSTAMPS

This section includes all the various pieces that have surfaced in limited quantity. In some cases, only one or two pieces with a certain backstamp have been found. Were these smaller companies, were their wares not as popular, or haven't I looked hard enough? Regardless, there are some very interesting items in this section. Hopefully more will surface. Also listed in this section are older Canadian pieces of Melamine, most marked Duraware (not to be confused with American Durawear).

Miscellaneous dinnerware. (Back row, left to right) A 6" tan mixing bowl marked "Duraware"; a 6" turquoise soup bowl marked "Duraware"; a tab-handled yellow soup bowl, 6", marked "Melmac Made in Canada." (Front row, left to right) Polystyrene cups in light green and yellow, marked "Duraware"; 5" and 6" bowls in yellow and turquoise, and pink and yellow cup and saucer sets, all marked "Table to Terrace Melmac Dinnerware."

Miscellaneous dinnerware. A forest green 8" divided serving bowl, and a 9.5" beige grill plate, both marked "Meladur by General American."

Miscellaneous dinnerware. (Back row) A pink soup bowl and 6" white underplate, marked "Lucent"; a 6" saucer in dark olive green, marked "The Plastic Potter"; a 9.5" turquoise plate marked "Malibuware." (Front Row) A tab-handled 4" bowl in green Polyethylene, marked "Victory of Chicago"; A 6" turquoise bowl marked "Lenoxware"; a pink cup and saucer set marked "Malibuware."

Miscellaneous dinnerware. (Back row) 10" dinner plates in turquoise and pink with white mottling, marked "Dailcyware by Home Decorators Inc." (Front row) A creamer and sugar (lid missing) in blue-green with red speckles, marked "Flite Lane Plastics Inc., St. Paul, Minn."; a 9" turquoise bowl marked "Apolloware by Alexander Barna."

Miscellaneous dinnerware. The bottom pieces are all marked
"Westinghouse" and hail from the Ovation or Newport lines.

Miscellaneous dinnerware. (Back row, left to right) Two beige bowls,
6" marked "Vollrath," and 7" marked "Ellingers Agatized Wood Inc.";
an 8" redwood bowl marked "Boltalite"; a 6" redwood bowl marked
"Halsey Inc." (Front row, left to right) A 4.5" ochre bowl marked
"Ellingers Agatized Wood Inc."; a low, mottled tan and white bowl,
5", marked "Plastiware — Pacific Plastic Products Company Los
Angeles, Calif."; a 6" ivory-colored bowl marked "Restraware." (Center) A creamer in institutional green, marked "U.S.M.C." on bottom. Also pictured is a 14" Polyethylene serving platter in mustard yellow, unmarked.

Miscellaneous dinnerware. (Back row, left to right) A 10" dinner
plate and a 6" bread and butter plate in institutional green, marked
"King Line"; a 9" plate in mottled dark green and yellow marked
"Devine Foods Inc."; a 10.5" tab-handled platter in mustard yellow, marked "Monte Carlo Melmac." (Front Row) Two 9" grill plates
in mustard yellow and mottled tan, marked "Trojan Ware USC."
Also pictured is a malt shop menu from the 1950s.

MISCELLANEOUS SOLID-COLORED DINNERWARE: WITHOUT BACKSTAMPS

This section contains all the pieces I have found that are unmarked — no backstamps, nothing! What a variety of pieces, too! Some are beautiful, high-quality, with interesting designs. Some are rather bizarre or poorly made — a possible explanation for the lack of a manufacturer's mark.

A number of these pieces may belong to marked sets. Sometimes not all the pieces in a set were marked — a saucer might carry the backstamp while the matching cup would not.

Miscellaneous table pieces, all un-marked. (Back row, left to right) A creamer in mustard yellow; a creamer and sugar in bubblegum pink; an identical creamer in turquoise; a small turquoise creamer. (Front row, left to right) A turquoise butter dish; a pink sugar bowl; a turquoise gravy boat. The sugar bowl in the front row has been seen in black.

Serving pieces, all unmarked. (Back row) 13" serving platters in sea green and pink. (Middle row) A 9" turquoise bowl. (Front row) Oval-shaped divided bowls, 10", in pink and in gray speckled with white.

Serving bowls, all unmarked, in copper-colored red with white speckles, yellow with white speckles, and turquoise.

Miscellaneous table pieces, all unmarked. (Top row, left to right) A cup and saucer in pink; a tab-handled bowl in turquoise; a bowl and underplate in yellow. (Bottom row) Three gravy boats from the same mold, in white with gray speckles, solid yellow, and turquoise with white speckles.

Miscellaneous cups, all unmarked. The gray cup (bottom right) is actually Spauldingware.

PATTERNED DINNERWARES

This is the Melmac that most people remember from their past — and have an aversion to. Every manufacturer of Melamine dinnerware produced patterned dishes after 1955, when the Melamine foil decal was first developed. The pattern selections are endless. Most are backstamped and easily identified. China and glassware companies ventured into patterned Melamine dinnerware in an attempt to cash in on the times. Their products were short lived and are a challenge to collect. The samples here are shown by subject: floral and fruit, leaf motifs, bold geometrics (very 1960s), and modernistic or contemporary. In the later 1950s some sets combined solid pieces with patterned ones. A set might have a patterned plate and saucer with a solid cup and bowl. Looking at old advertisements will help identify this trend.

Patterned dinnerware. (Left to right) A 6.5" plate marked "Boontonware"; a 15" platter and matching 7.5" plate with blue mums, unmarked.

Patterned dinnerware. (Left to right) A 10" plate marked "Style House"; a 7" plate marked "Miramar"; a 10" plate marked "Texasware."

Patterned dinnerware. (Left to right) A 10" plate marked "Beverly by Prolon"; a 7" plate decorated with birds of paradise, marked "Miramar Melmac"; a 10" plate marked "Apolloware."

Patterned dinnerware. (Left to right) a 10" plate, unmarked; a translucent white cup and saucer marked "Lenoxware"; a 10" plate, unmarked.

Patterned dinnerware. (Left to right) A 10" plate marked "Allied Chemical"; a 10" plate marked "Laguna Melmac"; A 10" plate marked "Park Avenue."

Patterned dinnerware. (Left to right) A 10" plate, unmarked; an 8" plate marked "Apolloware"; a 10" plate marked "Beverly by Prolon."

Patterned dinnerware. (Left to right) A 10" plate marked "Laguna Melmac"; a 6" saucer marked "Miramar Melmac"; a 10" plate marked "Sun Valley."

Patterned dinnerware. (Back Row, left to right) A 10" plate marked "Lenoxware"; a 9" plate, unmarked; a 10" plate marked "Durawear." (Front row, left to right) A 6" plate marked "Arrowhead"; an unmarked saucer; a 6" plate marked "Branchell"; an unmarked saucer.

Patterned dinnerware. (Left to right) A 10" plate marked "Lucent"; a 10" plate marked "Grant Crest"; 10" plate marked "Boontonware."

Patterned dinnerware with a stylized leaf design. (Left to right) a 6" saucer, a 10" plate, and an 8" plate.

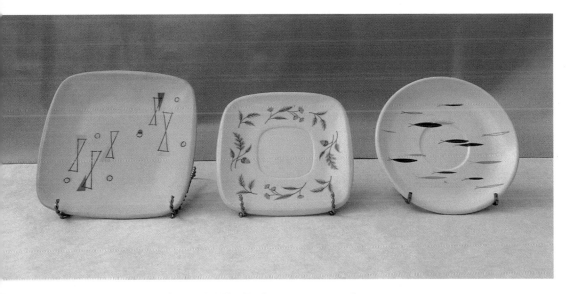

Patterned dinnerware. (Left to right) The Frolic pattern printed on a 7" plate in *Harmony House*'s Talk of the Town shape; a 6" saucer in *Harmony House*'s Talk of the Town shape; a 6" saucer marked "Watertown Lifetimeware."

How to Harm the Plastics Industry

The phrases 'quality control' and 'standards' plagued the post-war plastics industry. Retail and commercial customers were often subjected to false claims and misinformation coming from the plastics industry itself, and this became a significant problem. Because so much had developed so fast, the industry was forced to buckle down and set parameters. Some advice to those within the industry is nicely summed up in a 1947 *Modern Plastics Magazine* editorial entitled "How to Harm the Plastics Industry." Here is an edited version:

"A simple formula exists by the use of which any plastics company can do big and special damage to the plastics industry and to itself. Within the formula is a range of nasty techniques which may be used to produce the damage to this industry: Technique #1. Always refer to the plastics which the company uses as 'wonder' or 'miracle' material. Technique #2. Claim for these materials fantastic properties (three times as strong as steel) and imply that they offer properties never obtained in other materials. Technique #3. Produce a list of products and imply that the company has big and steady business in these items — and that the items are all successful. Technique #4. Lean heavily on experience with military applications, even if they have been purely developmental. Technique #5. Be generous in predictions on the future growth of these materials. Use out-of-context excerpts by figures in the industry to predict this fabulous future. Technique #6. Implicate endorsement and sponsorship by dropping names of big suppliers. Technique #7. Point up the company's extreme versatility, rather than its ability to concentrate. These techniques will cause people to waste time, money, and effort on 'dream' products based on a misunderstanding of the materials, and will bring later heartbreak to many who must inevitably find out the plain truth that many things for which a promise is implied cannot be done soundly with plastics."

Melamine dinnerware molded by Plastics Mfg. Co., has colorful molded-in design achieved through use of lithographed "foil"

Four-color plates courtesy American Cyanamid Co.

A place setting of Melamine dishes with modernistic designs, by Plastics Mfg. Company. From *Modern Plastics Magazine,* 1955.

Backstamps

TUPPER!

TUPPER! TRADE MARK

Bibliography

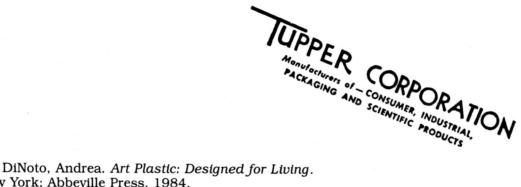

TUPPER CORPORATION
Manufacturers of – CONSUMER, INDUSTRIAL, PACKAGING AND SCIENTIFIC PRODUCTS

DiNoto, Andrea. *Art Plastic: Designed for Living.* New York; Abbeville Press, 1984.

Faulkner, Ziegfield, and Hill. *Art Today.* New York; Henry Holt & Co., 1940.

Lindenberger, Jan. *Collecting Plastics.* Atglen; Schiffer Publishing Ltd., 1991.

Lisfshey, Earl. *The Housewares Story.* Chicago; National Housewares Manufacturers Association, 1973.

McNulty, Lyndi. *A Price Guide to Plastic Collectibles.* Radnor; Wallace-Homestead Book Co., 1987.

Paul, Samuel. *How to Improve Your Home for Modern Living.* New York; H.S. Stuttman Co., 1955.

Wahlberg, Holly. *Everyday Elegance: 1950s Plastic Design.* Atglen; Schiffer Publishing Ltd., 1994.

————. *Modern Plastics Magazine.* New York; Modern Plastics, Inc., 1935-1960.

Value Guide

Please read this section carefully! The prices in this book are based upon the following factors: comparative price listings, prices observed at shops and sales, and availability and condition. If they seem on the high side remember this: *the prices listed in this book are for pieces in excellent or mint condition.* If this book lists a polystyrene butter dish for $8.00, that means no burns, cracks or scratches and with lid included. If you find the same butter dish with some damage or wear, do *not* pay the book price. The price ranges in this book are meant to accommodate the variance of prices across the country. A piece purchased in a secondhand shop in Arkansas will not carry the same price tag as an identical piece purchased in a Manhattan collectibles shop. Ultimately the final price you pay is between you and the seller.

Page	Pos	Description
96	T	l: platter 3-5; r: platter 5-8
	C	Back: platter 5-10; Front: trays 2-4 each
	B	l: Serving tray 8-12; server 5-8
97	T	t: metal tray 8-10; c: tray 4-6; b: tray 3-5
	B	l: dish 2-4; c: plate 3-5; r: dish 2-4
98	T	Back row l: salad set 10-12; r: salad set 8-10 Front row l: dishes 2-4 each; r: coaster/dish sets 3-5 each
	B	Back row l: tray 3-5; r: tray 5-8; Front row l: tray 3-5; r: tray 3-5
99	T	l: dispenser 3-5; c: dispenser 10-15; r: dispenser 10-20
	B	l: dispenser 4-6; c: shaker 3-5; r: shaker 3-5
100	T	l-r: sprinkler 5-8; dispenser 3-5; dispenser 5-8; sugar 3-5; sugar 4-6
	C	l-r: bear 8-12; bee 3-5; ketchup 2-4; mustard 2-4; mustard 2-4
	B	all dispensers 12-25 each
101	T	l-r: nut chopper 3-5; nut chopper 3-5; sugar 3-5; dispenser 4-8
	B	l-r: set 5-8; sugar 8-12; pitcher 10-15; set 4-6
102	T	l-r: salt and pepper 10-12 set; sugar 6-8; syrup 8-10
	C	l-r: sugar 10-15; Three sugars 5-10 each
	B	l-r: Three piece set 10-15; salt 4-6
103		clockwise from top: 5-8; 8-10; 8-10; 4-6; 6-8; 6-8
104	T	l: tray 5-8; c: tray 3-5; r: 3-5.
	BL	all slicers 4-8 each
	BR	all cake breakers 4-6 each
105	T	all strainers 8-12 each
	B	all knives 5-8 each
106	T	all knives 4-6 each
	C	all utility knives 7-10 each
	B	all utility knives 7-10 each
107	T	all carving pieces 7-10 each
	C	all carving pieces 7-10 each
	B	all carving pieces 7-10 each
108	T	all utility forks 7-10 each
	C	all knives 2-5 each
	B	l: set 8-12; c: set 15-30; r: set 8-12
109	T	all spoons 2-5 each
	C	all spoons 2-5 each
	B	all spatulas 4-6 each
110	T	all knives and spatulas 4-6 each
	C	all knives and spatulas 4-6 each
	B	all servers 5-10 each
111	T	all servers 5-10 each
	C	all spoons 4-6 each
	B	all ladles 4-6
112	T	all whisks 5-8 each
	C	all spatulas 5-8 each
	B	all mashers 10-15 each
113	T	all forks and knives 3-6 each
	C	l-r: spatula 5; knife 2-3; spoon 2-3; whip 2-4; whisk 2-3; brush 2
	B	l: slicers 3 each; knife 3-5; r: servers 2-3 each
114	TL	all utensils 2-5 each
	BR	l: vase 5-8; c: planter 4-6; planter 4-6; r: vase 3-5
115	T	l: planter 2-4; c: flower holder 3-6; r: flower pot 2-4
	C	l: planter 2-4; c: planter 2-4; r: planter 3-5
	B	all wall hangings 3-6 each
116	T	ivy design 8-12 each; rooster 2-4
	C	hangings w/planter boxes 5-8 ea.
	B	t-b: pierced basket 3-5; basket 2-4
117	T	l: basket 5-10; r: basket 4-6
	B	l: can 8-10; r: can 3-5
118	T	all cans 2-6 each
	B	l: can 3-6; r: can 4-8
119	T	cookie jars 15-25 each; sugar 6-8
	C	l: cookie jar 15-25; c: flour 10-15; r: canister 10-15
	B	Four piece set 20-30
120	T	Four piece set 25-35
	C	storage bin 8-10: salts 2-4 each; pepper 4-6; set 5-12
	B	bread boxes 15-25 each
121	T	pitchers 6-8; divided plates 5-8; cup 3-5
	B	napkin holder 4-6; basket 4-6; napkin holder 6-8; butter dish 4-8
122	T	soap dishes 3-5 each; wastebasket 5-10; flyswatter 2-5
	C	l: watering can 3-5; c: sprinkler 5-8; r: watering can 4-8
	B	trays 5-10 each
123	T	ring mold 6-8; tray 4-6; container 3-5
	C	pitchers 10, 8, 6, 5
124	T	Top row l-r: set 6-10; mug 3-5; creamer 3-5. Bottom row l: butter dish 3-6; c: covered dish; r: butter dish 4-6
	B	mixing bowl set, from largest down:6, 5, 5, 4, 3, 3
125	BL	all serving trays 2-5
	BR	l: candy dish 3-5; c: pitcher 2-4; r: candy dish 3-5
126	T	l: bowl 4-6; utensils 5 set; r: dish 3-5
	C	creamer and sugar 4-6 set; creamer 2-3; tray 3-5
	B	Back row l-r: holder 3; tumbler 2-3; vase 2-4; tumbler 2-3; holder 2 Front row l: butter dish 5; c: bowl 3; r: butter dish 5
127	T	6" tumblers, 3-5 each; 5.5" tumblers, 2-4 each; bowls 3-6 ea
	B	mugs 2-4 each
128	T	pitcher 6-8; 5.5" mugs, 3-5; 5" tumbler, 2-4; mugs 2-4; bowl 8-10
	C	5" tumblers, 2-4; bowls 3-5
	B	l-r: mug 3-5; cup 2-4; 3" mug, 2-4; 6" tumbler, 3-5
129	T	Top row: 5" tumblers, 3-5; 3.5" tumblers, 2-4. Bottom row: 3" mugs, 2-5
	C	l-r: mug 2-4; tumbler 3-5; pitcher 6-8; mug 3-5; mug 2-5
	B	l: bowls 3-5 ea.; r: tumblers 3-5 ea.
130	T	l: tumblers 2-4 ea.; r: jugs 5-8 ea.
	C	10" serving bowl 5-8; 6" bowls 3-5
	B	mugs 2-4 each; casserole 4-6
131	T	pitcher 8-12; 6" mugs, 4-6 each; bowls 3-5 each
	B	4" footed bowls 4-6 each
132	T	Salad bowl 8-10; utensil set 10-12; covered bowl 10-15; bowl 3-5
	B	tumblers 3-5; mugs 2-4; snack trays 6-8.
141	T	Back row: platter 3-5; oval platters 2-3 Middle row: oval bowls 2-4; 6" bowl, 2. Front row l-r: cup 2; bowl, 5", 2; cup, 2; bowl, 5.25", 3; mug, 3
	B	platter 3-5; bowl 2-4; oval platter 2-3; cup and saucer 3-4 set; creamer 2-3; 6" bowl 2; plate 3
142	B	Back row cup and saucers 4-6 set; divided serving bowl 10. Front row plates, 9.5" 4-8; 8" 3-6; 6" 3-4; bowls 3-6
143	T	Top row l: cup 2-3; plate 4-6; bowl 3-5. Bottom row l-r: cups 2-3 each; bowl 3-5; creamer 4-6.
	C	tray 5-10; bowl 2-4; dish 2-4; plate 4-6; 7" plate 2-4; 5.5" plate, 2-3
	B	Back row: divided bowl 3-5; platter 4-6; plate, 7" 2-3. Middle row: bowl, 6", 2-3; sugar and creamer 8 set. Front row: all cups and saucers 3-4 set
144	BL	sugar and creamer 8-12 set; plates, 10" 4-6, 9" 3-5, 8" 3-5, 7" 3-5, wide lip bowls, 6.75" 3-5; round lip bowls 2-5
145	T	Back row l: platter 5-8; r: platter 4-6. Front row butter dish 8-10; divided bowls 8-12
	C	all cups 3-5 each
	B	salt and pepper 10-15 set; cup and saucer 8-10 set. plates, 10", 5-8; 8", 4-6; 6", 3-5; bowls, 5.5" 3-5; plate, 6", 3-5
146	T	serving bowl, 14" 10-15; 6" plates 3-5
	C	serving bowl. 7" 4-6; bowls, 5" 3-5; cup and saucer 6-8 set; 6" plate 2-4; sugar bowl (lid missing), 5
	B	l-r: creamer 2-4; saucer 2-3; bowl, 2-4; all cups 2-3 each
147	B	l: 10.5" bowl, 4-8; 14" platter, 8-12; 12" platter, 8-10; ashtray 6-8
148	T	10" plate, 5-8; 7.5" patterned plate 3-5; saucers 2-5 each; 6" bowls, 4-6 each; cups and saucers 5-8 set
	C	Top row l: gravy boat 6-8; r: butter dish 8-12. Bottom row l: creamer and sugar 10-12 set; creamer 5-7
	B	6" bowls 4-6 each; Butter dishes 8-12 each; 12" salad bowl, 12-15; salad spoon 6-10
149	T	10" plate, 5-8; 7.5" plate, 4-6; 6" plate, 3-4; 12" platter, 8-10; cup & saucer 5-8; creamer & sugar 10-12; 3.5" tumbler, 4-6
	C	plates, 10" 5-8; plates, 7.5" 4-6; bowls, 5" 3-5; cup & saucers 5-8 set; creamer 5-7
	B	5" tumbler, 5-8; 6" plate, 3-4; 10" plate, 5-8; 7.5" plates, 4-6 each; 3.5" tumbler, 4-6
150	T	10" bowl, 8-10; 7.5" bowl, 3-5; 8" bowl, 4-6; creamer and sugar 8-10 set; cup & saucer 5-8
	C	all serving bowls 6-8 each
	B	platter 8-10; all creamers and sugars 3-5 each
151	T	plates, 10" 5-8 cups & saucers 4-6 set
	B	Back row 7.5" bowls, 3-6; 5.5" bowl, 2-5; cup & saucer 4-6 set. Front row all plates 3-5
152	T	plate 6-8; platter 8-10 shaker 3-5; cup and saucer 3-5; serving bowl 5-7; plate 3-5
	B	platter 8-10
153	T	bowls 2-4 each; all mugs 2-4; all divided plates 3-6
	B	9" plates, 3-5; 6" plate, 2-4; 4.5" bowl, 2-4; all coasters 2-3 ea.
154	T	all cups and saucers 3-6 set; plate 2-4
	B	Back row l: serving bowl 6-10; r: 10" plate, 5-8. Front row l: cup and saucers 3-5 set; sugar & creamer 8-12 set
155	T	serving bowl 6-8; 10" plates, 5-8

	B	all 6" plates 2-4 each
156	T	Top row: platters 6-8; creamers 3-4 each; sugar 3-5, mug 2-4. Bottom row: all bowls 2-4 each; saucer 2-3
	B	10" plates, 3-6; cup and saucer 4-6 set; 5" bowl, 2-3; 7" bowl, 3-4; 8" plate, 3-5
157	T	Top row l-r: mug 3-5; mug 3-5; two-handled mug 8-12. Bottom row l-r: bowl 2-4; two-handled mug 2-4; tumbler 1-2; tumbler 2-5
	B	all divided plates 6-10 ea.; serving bowl 3-5; divided bowl 4-6
158	T	Back row l: 9.5" plate, 4-5; 8" plate, 2-4. Middle row : all 7" plates, 3-5 ea. Front row: all coasters 2-3 each
	B	10" plates, 4-6; 5" tumbler, 3-5; 3.5" tumbler, 3-4; creamer 3-5; 6" bowl, 3-5; cup 2-5; low bowl 2-4
159	T	10" plates, 3-6; 6" plate, 2-4; 6" bowl, 2-5; cup and saucers 4-6 set
	B	Serving bowl 3-5; creamer 4; saucers 2-3 each; plates 2-4 each; bowls 3-4 each
160	B	all cups 3-5 each
161	T	serving bowl 3-6; creamer and sugar 6-10 set; butter dish 10-12; 10" plate, 4-6; 8" plate, 3-5; tab-handled bowl 6; cup and saucer 4-8; 7" bowl, 5-8.
	B	serving platter 10-12; pitcher 12-18; creamer and sugar 8-12; serving bowl 4-8; bowls 3-5; cup and saucer 5-9
162	T	platter 4-6; sugar 3-5; creamer 3-5; serving bowl 4-6; bowl 2-3; plate 2-3
	C	Back: plates 3-5 each; tumblers 5-8 each; cups and saucers 3-5 set
	B	all tab handled bowls 3-5 each
163	T	creamer 2-4; sugar 2-4; bowl, 6" 2-3; bowls 2-4 each; serving bowl, 3-5; cup and saucer 3-5
	C	platter 3-6; plates 3-4 each; bowls 2-5 each
	B	serving bowl 4-8; creamer & sugar 4-6, cup & saucers 3-4 set
164	T	sugar bowls 2-3 ea.; cup & saucers 2-4 set; serving bowl 3-5
	B	bowls 2-4 each; 10" plate, 3-5; 6" plate, 2-3; platter 4-6; cup and saucers 3-6 set
165	C	Back row l-r: 6" plate, m1-2; bowl 1-2; serving bowl 2-4; creamer 2-3; 7.5" plate, 2-3. Front row l-r: cup and saucer 2-3

		set; 6" bowl, 1-2; gravy boat with underplate 3-4
166	T	l: sugar 4-6; c: creamer 4-6; r: sugar 4-6.
	C	Back row l-r: sugar and creamer 8-12 set; plate 3-6; serving bowl 8-10. Front row l-r: cup and saucer 3-6 set; 10" plate, 3-6; 6" plate, 2-4; bowl 2-4; bowl 2-4 Inset t-b: platter 6-10; 10" plate 5-8; creamer 4-6; 6" plate, 3-5; 10" plate, 5-8
	B	all 10" plates, 3-6 ea.; 6.5" plates, 2-4; cup & saucers 3-6 set
167	T	platters 4-6 each; plate 3-6; all bowls 3-4 each; creamer 3-4
	C	10" plates, 6-10 each; cup and saucers 8-10 set
	B	14" platter, 12-15; all serving bowls 8-12 each; plate 4-8; 12" platter, 10-12
168	T	all 10" plates, 3-6 each; bowls 2-4 each; cup & saucers 3-6 set
	B	serving bowls 3-6 each; cup and saucers 2-5 set
169	B	7.5" plates 4-6 each; 10" plate 6-8; serving bowl 12-15; cups 3-5
170	TR	serving bowl 4-6; plates 2-3 ea.; cup & saucers 4-6 set
	C	Top row: creamer & sugar 5-8 set; platter 4-6; serving bowl 4-8 Bottom row l-r: creamer and sugar 5-8 set; cups 2-3 each
	B	all plates 3-4 each; tab-handled bowl 2-3
171	T	platter 3-6; cup and saucers 3-5 set; gravy boat 5; sugar 5; mug 3-4; butter dish 6-8
	C	platter 4-8; 7" plate, 3-4; sugar and creamer 6-10 set; plates 3-4 each; bowls 2-3 each
	B	Back row l: serving bowl 4-8; r: platter 4-8. Front row l-r: cup and saucer 3-5 set; sugar and creamer 6-10 set; plates 3-4 each; bowls 3-4 each
172	BL	7" plates, 3-4 each; low bowl 3-5; saucers 2-3 each
173	T	Back row l: Divided plate 3-5; c: mugs 3 each; r: plate, 3-4. Front row l: serving dish 3-4; r: low bowl 2-3; bowl 2-3
	C	Back row platter 3-6; plates 2-4; saucers 1-2. Front row cup and saucer 2-4; 6" plate, 1-2; 5" bowl, 1-2; mixing bowl 5-8; serving bowl 3-4; 10" plate, 3-4; mug 3; sugar bowl 3-5
	B	Back row platters 3-6 each; plate 1-2; Center row: creamer and

		sugar 6-8 set; bowl 1-2; cup and saucer 2-4. Front row serving bowl 3-4; butter dish 4-6
174	T	serving bowl 4-6; cup and saucer 2-5 set; cups 3 each; sugar & creamer 5-8 set; sugar 3-4
	C	10" plates, 3-5; serving bowl 4-6; cup 3; plate 2-3; 6" bowl, 2-3; 5" bowl, 1-2
	B	Back row l-r: mixing bowl 6; 6" bowl, 4; tab-handled bowl 5. Front row l-r: cups 3 each; bowls 3-4 each; cup & saucers 5 set
175	TL	grill plate 8-12; serving bowl 10-15
	TR	Back row bowl 3; underplate 3; saucer 2; plate 4-5. Front row: tab-handled bowl 3; bowl 3; cup and saucer 5 set.
	B	Back row: plates 3-5 each. Front row: creamer and sugar 6-10 set; bowl 5-8
176	T	Various cups and creamers 2-5 ea.
	C	Back row l-r: 6" and 7" bowls, 2-4 each; 8" bowl, 4; 6" bowl, 3 Front row l-r: 4.5" bowl, 3; low 5" bowl, 2; 6" bowl, 3. Center: creamer 5
	B	Back row l-r: 10" plate, 4; 6" plate, 2; 9" plate, 5; tab-handled platter, 6. Front row: grill plates 3-6 each
177	T	Back row l-r: creamer 3; creamer and sugar 8 set; creamer 4; creamer 3. Front row l: butter dish 5-7; c: sugar bowl 4; r: gravy boat 5
	C	Back row: platters 3-5 Center row: bowl 4. Front row: bowls 5 ea.
	B	all serving bowls 4-8 each
178	T	Top row l: Cup and saucer 5; c: tab-handled bowl 5; r: bowl and underplate 5 set. Bottom row: all gravy boats 5 ea.
	C	all cups 2-4 each
	B	l: plate 2; c: platter 4-5; r: plate 3
179	T	l: plate 3; c: plate 2; r: plate 3
	C	l: plate 4; c: plate 3, r: plate 3
	B	l: plate 3; c: cup and saucer 4 set; plate 3
	R	t: plate 3; c: plate 3; b: plate 3
180	T	l: plate 3; c: plate 2; r: plate 3
	C	l: plate 4-5; c: saucer 3; r: plate 4-5
	B	Back row: l: plate 3; c: plate 3; r: plate 3. Front row l-r: plate 3; saucer 2; plate 2; saucer 2
181	T	l: plate 4; c: plate 3; r: plate 4
	C	l: saucer 2; c: plate 4; r: plate 3
	B	l: plate 4; c: saucer 3; r: saucer 3

Notes

Notes

See Catalin pictures
20-21 Barware
26 shaker pitcher
36 youth set   15 each

Barware
26 shaker pitcher